ANATOMY
OF THE MOVING BODY

ANATOMY
OF THE MOVING BODY

A BASIC COURSE IN BONES, MUSCLES, AND JOINTS

Theodore Dimon, Jr.

with illustrations by
Megan Day

North Atlantic Books
Berkeley, California

Printed in the United States of America

North Atlantic Books are available through most bookstores. To contact North Atlantic directly, call 800-337-2665 or visit our website at www.northatlanticbooks.com. Substantial discounts on bulk quantities of North Atlantic books are available to corporations, professional associations, and other organizations. For details and discount information, contact the special sales department at North Atlantic Books.

Published by North Atlantic Books
P.O. Box 12327
Berkeley, California 94712
Distributed to the book trade by Publishers Group West

Cover design by Megan Day
Book design by Jan Camp

Anatomy of the Moving Body is sponsored by the Society for the Study of Native Arts and Sciences, a nonprofit educational corporation whose goals are to develop an educational and crosscultural perspective linking various scientific, social, and artistic fields; to nurture a holistic view of the arts, sciences, humanities, and healing; and to publish and distribute literature on the relationship of mind, body, and nature.

Library of Congress Cataloging-in-Publication Data

Theodore, Dimon.
    Anatomy of the Moving Body : a basic course in bones, muscles, and joints /
by Theodore Dimon, Jr.
    p. cm.
  Includes bibliographical references and index.
  ISBN 1-55643-207-0 (alk. paper)
  1. Musculoskeletal system—Anatomy. 2. Human locomotion. 3. Alexander technique.
I. Title

qm100 .d56 2001
611'.7—dc21
CIP

00-048028

1 2 3 4 5 6 7 8 9 / 05 04 03 02 01

To my grandfather, Panos Dimon, with unbounded love

TABLE OF CONTENTS

List of Illustrations   ix

Preface   xi

Introduction

What is Anatomy?   3

   1. Anatomical Terminology   7

   2. Bones, Muscles, and Joints   15

Head and Neck Region

   3. The Skull   23

   4. Base of the Skull and Its Attachments   27

   5. Muscles of the Face and Jaw   37

   6. Suspensory Muscles of the Larynx   43

   7. The Tongue   47

   8. Muscles of the Palate   51

   9. Muscles of the Throat   57

   10. The Larynx   61

Spine and Trunk Region

   11. Anterior Muscles of the Cervical Spine   69

   12. The Vertebrae of the Spine   73

   13. The Spine and Its Supporting Ligaments   83

   14. Muscles of the Back: Deep Layers   89

   15. Muscles of the Back: Middle and Superficial Layers   99

   16. Muscles Attaching to the Front of the Spine   107

Contents

Thorax and Abdomen

17. The Thorax and Muscles of Respiration   113

18. The Abdominal Muscles   125

19. Suspensory Muscles of the Thorax   131

20. The Spiral Musculature of the Trunk   137

Shoulder Girdle and Upper Limb

21. The Shoulder Girdle   143

22. Muscles of the Arm and Shoulder   153

23. Muscles of the Forearm   161

24. Muscles of the Hand and Wrist   167

25. Intrinsic Muscles of the Hand   179

Pelvis and Lower Limb

26. The Pelvic Girdle   187

27. Muscles of the Pelvis and Hip   199

28. Muscles of the Thigh   207

29. The Knee, Lower Leg, and Ankle   217

30. Muscles of the Ankle and Foot   227

31. Intrinsic Muscles of the Foot   239

Index   249

ILLUSTRATIONS

Fig. 1.   Anatomical planes

Fig. 2.   Anatomical directions

Fig. 3.   The skull

Fig. 4.   The base of the skull

Fig. 5.   Flexors and extensors
           attaching to base of skull

Fig. 6.   Muscles supporting hyoid
           bone and larynx

Fig. 7.   Base of the skull and
           muscles of the throat

Fig. 8.   Muscles and joint of jaw

Fig. 9.   Muscles of facial expression

Fig. 10.   Muscles of the jaw

Fig. 11.   Suspensory muscles of the larynx

Fig. 12.   Suspensory muscles of
           the larynx (cont.)

Fig. 13.   The tongue

Fig. 14.   Muscles on the floor of mouth

Fig. 15.   Muscles of palate

Fig. 16.   Muscles of the throat

Fig. 17.   The pharynx

Fig. 18.   The larynx

Fig. 19.   Intrinsic muscles of the larynx

Fig. 20.   Anterior muscles
           of cervical spine

Fig. 21.   Vertebral column

Fig. 22.   The vertebrae and spine

Fig. 23.   Atlas and axis (C1 and C2)

Fig. 24.   The skull and head/neck joints

Fig. 25.   Ligaments of the spine

Fig. 26.   Lower spine showing
           pinched disc

Fig. 27.   Back muscles: 1st layer
           (transversospinalis muscles)

Fig. 28.   Back muscles: 1st layer (cont.)

Fig. 29.   The sub-occipital muscles

Fig. 30.   Back muscles: 2nd layer
           (sacrospinalis or erector spinae)

Fig. 31.   Back muscles: 3rd layer

Fig. 32.   Back muscles: 4th layer

Fig. 33.   Back muscles: 5th
           (superficial) layer

Fig. 34.   Muscles attaching to the
           front of the spine

Fig. 35.   The rib cage

Fig. 36.   The costovertebral joints

Fig. 37.   Ribs during exhalation and
           inhalation

Fig. 38.   The intercostal muscles

Fig. 39.   Transversus thoracis

Fig. 40.   The diaphragm

Fig. 41.   The abdominal muscles

Fig. 42.   Rectus abdominis muscle

Fig. 43.   The scalene muscles

ix

x

Fig. 44. Suspensory muscles of the thorax

Fig. 45. Muscles of the thorax (cont.)

Fig. 46. Spiral musculature of the trunk

Fig. 47. Joints of shoulder girdle

Fig. 48. Scapula and shoulder joint

Fig. 49. Trapezius, teres major,
          and latissimus dorsi

Fig. 50. Scapula muscles

Fig. 51. Serratus anterior
          and pectoral muscles

Fig. 52. The rotator cuff muscles

Fig. 53. The deltoid muscle

Fig. 54. Flexors of the arm

Fig. 55. Triceps brachii muscle

Fig. 56. Bones of elbow and forearm

Fig. 57. Supinators and pronators
          of the forearm

Fig. 58. Bones of wrist and hand

Fig. 59. Joints of the wrist

Fig. 60. Joints of the thumb

Fig. 61. Extensors and flexors of wrist

Fig. 62. Flexors of digits

Fig. 63. Extensors of digits

Fig. 64. Intrinsic muscles of the thumb

Fig. 65. Intrinsic muscles of the little finger

Fig. 66. Interossei and lumbricales

Fig. 67. The pelvis, the right innominate
          bone

Fig. 68. Landmarks of the pelvis

Fig. 69. The hip joint and femur

Fig. 70. Ligaments of the pelvis

Fig. 71. Ligaments of the hip joint

Fig. 72. The iliopsoas muscle

Fig. 73. The pelvic diaphragm

Fig. 74. The deep muscles of the hip

Fig. 75. The gluteals

Fig. 76. The adductors

Fig. 77. Muscles of the thigh

Fig. 78. The quadriceps muscles

Fig. 79. The hamstring muscles

Fig. 80. The knee joint

Fig. 81. Bones of lower leg

Fig. 82. The ankle joint

Fig. 83. Ligaments of the ankle

Fig. 84. Bones of foot

Fig. 85. Joints of the foot

Fig. 86. Anterior muscles of the leg

Fig. 87. Lateral muscles of the leg
          (peroneal muscles)

Fig. 88. Muscles on the back of the leg

Fig. 89. Muscles on the back
          of the leg (cont.)

Fig. 90. Intrinsic extensors of the foot

Fig. 91. Interossei muscles

Fig. 92. Intrinsic muscles
          of the little toe

Fig. 93. Intrinsic muscles of the big toe

Fig. 94. Intrinsic flexors of the toes

Fig. 95. Arches of the foot

PREFACE

This book was originally written as a series of lectures for a basic course in anatomy given at the Dimon Institute for the Alexander Technique in Cambridge, Massachusetts. Its purpose is to provide teachers and students of movement with a basic text covering all the muscles, bones, and joints relating to movement. More specifically, it is designed for movement educators who are putting together their own courses on anatomy and require a basic manual to work from that provides, not just drawings and names of anatomical structures, but written lectures which tie this material together into a coherent series of presentations. The lectures also provide sufficient explanation to enable the book to serve as a "self-help" manual for students of movement and dance.

The overwhelming consideration in putting together this book—and the feature, I think, that distinguishes it from many others on the subject—has been simplicity and clarity of presentation. Anatomy as a subject matter has developed in close association with medical science; as such, anatomical texts tend to be highly technical and detailed, and often seem designed more to intimidate and impress than to enlighten. For many, such books not only fail to convey the necessary and relevant information about anatomy as it pertains to movement; they also contain much that is irrelevant and unnecessary to such study. In the process of giving too much detail, they also fail to explain how muscles and bones work in simple terms, and so further obfuscate the real issue for students and teachers of movement, which is not merely to know, but to understand.

This book, accordingly, contains only what is really needed by the student of movement—with accompanying illustrations—and leaves out many of the complex details that are not necessary for the present purposes. For similar reasons, the illustrations do not show layer upon layer of muscle, but only one or several at a time. The written talks provide necessary background information which is essential for students to assimilate this information. Without written explanations about anatomical structures and how we are designed to move, it is difficult to retain and make sense of what we are viewing.

The book is divided into sections presenting muscles and joints in the simplest possible manner, while not oversimplifying or leaving out necessary detail. The book contains 31 chapters designed to be presented weekly in a half-year or one-year course. Each chapter covers a basic region of the body and is accompanied by simple and clear illustrations. The book begins with two chapters on basic terminology and anatomical structures followed by separate chapters covering different regions of the body. These include the head and neck region, the back, abdominal muscles, and so on.

The book also includes information about bones; origins and attachments of muscles and related actions; joints, major ligaments, and actions at joints; discussion of major functional structures such as the pelvis, shoulder girdle, ankle, and hand; etymology of anatomical terms; major landmarks and human topography; and structures relating to breathing and vocalization.

Notice that a good deal of attention is given to the larynx and muscles of the face and throat, as well as to the bones of the skull. Although these structures do not relate directly to overt or external movement, they are an integral part of the musculoskeletal system and are therefore an essential part of what any dancer, performer, or movement educator must know about the body and how it works. As with other systems addressed by the book, these muscles are grouped by region.

My hope is that teachers and students of movement will find this text useful in increasing not only an intellectual understanding of anatomy but also self-awareness of the body in action. With this end in view, the present book will be followed by a second volume focusing more directly on human movement in relation to anatomical design and function.

INTRODUCTION

What is Anatomy?

For most people, anatomy is somewhat intimidating. The inside of the body, the "scientific" nature of the subject, the mysterious structures involved, and the complex names all make anatomy a somewhat frightening, not to mention boring, subject. But anatomy can be not only fun but outright fascinating; and there is little that is really intimidating once you sort through some of the terminology.

"Anatomy," if you look up the root of the word, means "to cut up." Put simply, it is physical description of the parts and structures of the body, which people have learned about by cutting up the body. In contrast, "physiology" means "study or discourse on nature"—in other words, how the body functions. Anatomy is structure—the identifying and naming of parts. Physiology is function—understanding how things work.

I should note from the outset that the following lectures do not cover general anatomy, which includes the heart and vascular system, the digestive and reproductive systems, the nervous system, and so on. The kind of anatomy we're concerned with here is the musculoskeletal system—that is, structures that relate to movement and motor activity, including respiration and vocalization. In particular, these lectures are for people who have a practical interest in muscle tension, awareness, physical control, and movement.

Of particular interest to the study of movement and muscles is functional anatomy—looking at the body from the point of view of how we function. For instance, when we examine how the legs in humans have become modified to allow us to stand upright, as compared with those of a four-footed animal, this is anatomy from a functional point of view. This kind of anatomy is helpful if you are trying to understand in a practical way why

4

we are designed the way we are; it is also helpful if you are teaching others. These lectures include a large dose of functional anatomy, since those of us who work with students in a practical way are concerned not simply with where things are, but how the body works in balance and movement. Nevertheless, any serious professional in the field of movement and functioning requires a basic background in traditional anatomy—identifying the major muscles, bones, and joints—as the basis for further study. Providing a grounding in traditional anatomy from the point of view of movement and functioning is the primary goal of these lectures.

A word about terminology. When you first become acquainted with joints, ligaments, and muscles, the anatomical terminology used to name these parts often seems long and intimidating: atlanto-occipital joint, sternocleidomastoid muscle, and so on. But remember, first of all, that anatomy as we know it today is based on Latin and Greek; when you identify the roots of some of the words ("cranium" means "a helmet" in Greek; "acetabulum" was a Latin word for "a small bowl"), they don't seem so intimidating. Most anatomical structures, in fact, were named according to what they resembled; there is even a structure (*os innominatum,* meaning "no name") that was given its name because it didn't resemble any known object! There is no mysterious terminology or language that you are required to know in order to understand anatomical terms; they were, and are, simple descriptions in other languages.

Secondly, the names for many muscles and joints are long because they identify the two points they connect. For instance, "atlanto-occipital" refers to the joint formed by the atlas and the occiput; the "sternocleidomastoid" muscle attaches from the sternum and clavicle to the mastoid process of the head. In many cases, if you know the two points a muscle connects, you know the muscle. In other cases, the names are not medical terms belonging to the special province of doctors and clinicians, but simply descriptive names in another language which generally only medical people use. Knowing this helps to make sense of what otherwise seem to be long and technical names.

Finally, do not expect, or try, to learn all the material contained in this book at once. As with most subjects, understanding what you are learning takes time. As you gain experience in the field, scattered bits of information begin to fit into a larger picture. You will find that particular muscles and joints have become familiar to you, that you have become fluent with the anatomical names for these structures, and that as a student, teacher, and professional you can increasingly draw upon your growing repertoire of anatomical knowledge with confidence and ease.

5

1. Anatomical Terminology

A set of terms has evolved to describe spatial positions and relationships in the human body when speaking of anatomy or movement. They are all related to the standard anatomical position, which is standing erect, palms of the hand forward, as in most anatomy charts.

Planes

In order to provide a means for describing where anatomical structures are located three-dimensionally, the body is divided into three planes (Fig. 1).

a. Median, or saggital plane
    The vertical plane dividing the body into left and right halves.

b. Coronal, or frontal plane
    The vertical plane dividing the body into front and back halves. The terms anterior/posterior relate to this plane. Some writers use this term to describe the point at which the head balances on the spine.

c. Horizontal, or transverse plane
    The horizontal plane dividing the body into upper (cranial) and lower (caudal) parts.

Anatomical Directions and Positions (Fig. 2)

a. Superior, or cranial
    Above, or towards the head.

Fig. 1. Anatomical planes

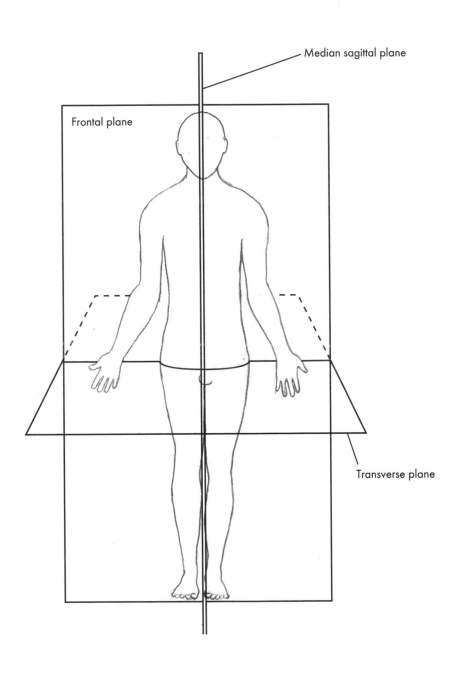

b. Inferior, or caudal
    Below, or towards the feet.

c. Anterior, or ventral
    Before, or nearer the front (corresponds to frontal plane).

d. Posterior, dorsal
    Behind, or towards the back (corresponds to frontal plane).

e. Medial
    Lying closer to the median plane, or towards the center from the side.

f. Lateral
    Lying further away from the medial plane.

g. Proximal
    (Used only with limbs) Closer to the root of the limb.

h. Distal
    (Used only with limbs) Closer to the end of the limb.

Types of Movement

Flexion
To bend at a joint; from the Latin *flexus,* to bend. When you flex at a joint, the angle at the joint diminishes.

Fig. 2. Anatomical directions

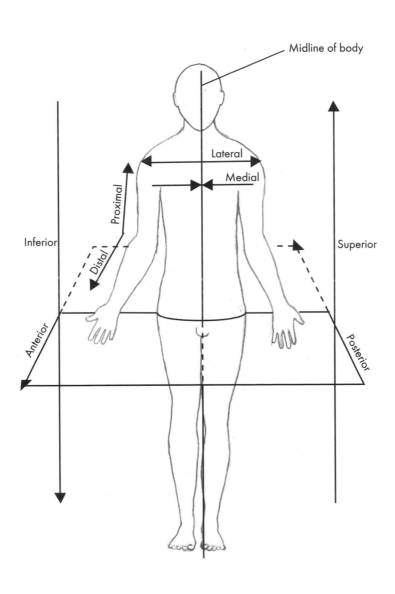

Extension

To stretch out, or straighten at a joint; from the Latin *"ex,"* which means "out," and *"tendere,"* which means "to stretch"—in other words, to stretch out. When you extend at a joint, the angle at the joint increases.

Hyperextension

To straighten at a joint beyond 180 degrees; to overextend.

Abduction (from the Latin, meaning "to lead away")

To move away from the midline or saggital plane.

Adduction (from the Latin, meaning "to lead towards")

To move towards the midline.

Circumduction

A circular movement of the trunk or limb that combines flexion, abduction, extension, and adduction.

Lateral flexion

To bend the trunk laterally or sideways.

Rotation

To rotate the body or a limb around its long vertical axis.

Plantar flexion

To bend at the ankle by bringing the toes away from the knee.

Dorsiflexion
To bend at the ankle by bringing the toes towards the knee.

Inversion
Bending the foot inward.

Eversion
Bending the foot outward.

Pronation
Rotation of forearm so that the palm turns downward.

Supination
Rotation of forearm so that the palm turns upward.

Ulnar deviation
Sideways movement of the hand at the wrist.

Radial deviation
Return movement from ulnar deviation.

Cortical opposition
The movement of the thumb that opposes the fingers.

Areas of Anatomy

Comparative anatomy: anatomy based on comparison of different species or groups.

Embryology, or developmental anatomy: the study of the human organism in its immature condition from fertilization of the ovum to birth.

Gross anatomy: anatomy that deals with structures visible to the naked eye.

Histology: the study of bodily cells and tissues.

Kinesiology: the scientific study of human movement.

Morphology: the study of the form and structure of animals and plants.

Neuroanatomy: anatomy of the nervous system, including the brain.

Other Related Terms

Agonists, or prime movers: the muscles that bring about movement.

Antagonists: the muscles that oppose the agonists, or move the bone in the opposite direction. Many muscles or muscle groups have opposing muscles—i.e., the biceps, which are opposed by the triceps; and the extensors of the back, which are opposed by the flexors.

Appendicular skeleton: the skeleton of the limbs.

Articular surface: the smooth area on a bone which articulates with another bone to form a joint.

Axial skeleton: the skeleton of the head and trunk.

Condyle (*kondylos,* a knuckle): the rounded knuckles at the end of long bones which form points of attachment for muscles and points of articulation for joints.

Crest: large ridge on a bone.

Foramen: a hole or opening in a bone.

Lip: borders of the crest.

Ontogeny: the lifespan of the organism from fertilization to mature adult.

Phylogeny: the evolutionary history of the group (or phylum) to which the organism belongs.

Process: localized projection.

ROM: abbreviation for range of motion.

Spine: a prominent ridge.

Tubercle/tuberosity: rounded lump of bone.

2. Bones, Muscles, and Joints

In the study of musculoskeletal anatomy, the main focus is on muscles and the bones to which they attach. Bones also connect to each other, forming joints. So the three main structures we must deal with in musculoskeletal anatomy are bones, joints, and muscles.

Bones, Joints, and Ligaments

Bones form the framework for the body. They also serve as levers that are acted upon by muscles. Bones come in varied shapes and sizes. Long bones are found in the limbs, where they act as levers for support and locomotion. Short bones function for strength and compactness. Flat bones have a protective function or provide broad surfaces for muscular attachments (e.g., the skull and shoulder blades).

As mentioned, joints are areas where bones are linked together. Some joints, such as the sacroiliac joint, are very inflexible and capable of little or no movement; the bones are simply linked together by fibrocartilage and reinforced by ligaments. Other joints are freely movable, permitting the bones to form levers that hinge or pivot with one another; these are called synovial joints because they contain a synovial fluid which lubricates the articulating surfaces, allowing them to glide or move against one another. (Another name for movable joints is *diarthroses,* which is related to the word "arthritis.") There are several kinds of synovial joints such as hinge joints, ball-and-socket joints, gliding joints, pivot joints, and saddle joints; we'll see examples of these as we proceed.

Bones do not actually rub against each other where they articulate. The articulating surfaces are covered by cartilage—a tough, smooth, and shiny fibrous material that helps to protect the bone and allows movement at the joint. Cartilage, which is Latin for "gristle," can be seen on the freshly cut bones of meat bought at the butcher shop; you can also find it on the joints of real skeletons, which often retain some of the cartilage lining. Cartilage absorbs pressure, reduces friction, and protects the bone. It also tends to increase the articulating area of a joint and absorbs the fluids that lubricate the joint, helping to keep the fluid from dissipating. Arthritis occurs when this cartilage is damaged or worn away with age or constant pressure, which then causes damage to the bones, which rub together.

(There are actually two kinds of arthritis. When the cartilage wears away, this is called osteoarthritis. Rheumatoid arthritis is inflammation of the synovial linings of the joints which eventually causes damage to the cartilage. The cause of rheumatoid arthritis is not clear. Some people think it is a virus; possibly other factors such as tension and emotional states enter into it. Unlike osteoarthritis, which occurs with wear and tear, rheumatoid arthritis can occur spontaneously at any age.)

The lubricating fluid in a joint is called synovial fluid. This fluid protects the articulating surfaces of the bones, which would otherwise succumb to friction and wear away. Joints are usually enclosed within a sleevelike structure, or joint capsule. This capsule binds together the bones and retains the synovial fluid; it also contains a membrane on its inner layer which secretes the synovial fluid. This fluid is depleted during activity, but is replenished during rest.

Joints are bound together by ligaments, a word that means "to bind." They are made up of very strong fibers. All the primary joints are firmly bound together by ligaments: hips, head/neck joint, shoulder joints, wrists, ankles, fingers, toes, vertebrae of the spine, and so on. Ligaments are quite tough and usually non-elastic, although there are some ligaments, which are yellow in color, that are elastic. However, even non-elastic ligaments can become stretched from too much stress, as when we slump, which takes the muscles out of gear and forces us to rely too heavily on ligaments for support. Ligaments cannot contract, but they do have a limited number of sensory nerves. Torn ligaments result from undue stress on joints, with knee and ankle injuries being the most common.

Muscles, Fascia, and Tendons

Muscles are attached to bones, and by contracting they produce movement. So bones function as levers, and muscles as motors that move the levers. Because muscles connect to bones, they are considered a form of connective tissue. Muscles taper to become tendons, attaching via the tendon to bones and cartilage. So tendons connect muscle to bone.

Fascia, which means "bandage," are tendinous fibers that connect the skin and underlying structures and form sheaths for muscles, binding their fibers into one unit. Fascia are also a form of connective tissue and can be found throughout the body.

Types of muscle

a. Striated or skeletal muscle attaches to bone and is capable of producing or assisting in movement. This type of muscle is also referred to as voluntary muscle.

b. Smooth or involuntary muscle can be found lining the intestines and stomach.

c. Cardiac muscle serves the specialized function of pumping blood.

Of these three types of muscle, the one that directly concerns us is striated or voluntary muscle.

Attachments

Skeletal muscles are attached from one bone to another. When a muscle contracts to produce movement, one of its attachments remains relatively stationary while the other moves. The larger, more stable structure is considered the *origin* of the muscle; the smaller, less stable structure, to which the muscle attaches, is called the *insertion*.

Muscles have different kinds of origins and insertions. Some muscles have a very broad origin. Some muscles attach directly to bones; other muscles taper into long tendons that insert into bones. Some muscles have several origins, or heads, which then converge into one insertion.

Muscular forms

Muscles have different forms and fiber arrangements, depending on their function. In the limbs they tend to be long; in the trunk they tend to be broader and to form sheets that wrap around the body. Muscles that stabilize parts of the body, such as those found in the hip, are short and squat. Muscles that produce large, sweeping movements, such as the muscles of the legs and arms, are longer and thinner, and because of their greater length, can contract more and are therefore capable of producing greater movement than short muscles. Sartorius, one of the leg muscles, is nearly two feet long, whereas the muscles of the eye are minute in size.

Here are some of the different fiber arrangements found in muscles:

1. Fusiform or spindle: rounded muscle that tapers at either end.

2. Quadrilateral: flat and four-sided.

3. Penniform, rhomboidal, or feather-like: fibers that extend diagonally from a long tendon, giving the appearance of one side of a feather.

4. Bipenniform: a double penniform muscle.

5. Triangular or fan-shaped: flat muscle that fans out from a narrow attachment at one end.

6. Sheet: flat muscles that span out over large areas, as in the trapezius or latissimus dorsi muscle.

19

Muscles are composed of bundles of fibers held together by very thin membranes. Within the fibers are thousands of tiny filaments, which slide along each other when the muscle is stimulated by a nerve. This causes the muscle to shorten, or contract.

Muscles that produce movement are called *agonists;* muscles that produce the opposite movement are called *antagonists.* When a muscle shortens in length, this is called *isotonic* contraction. When it contracts but cannot overcome the resistance of weight, immovable objects, or the opposing action of the antagonistic muscles, this is called *isometric* contraction.

HEAD AND NECK REGION

3. The Skull

The skull is composed of two parts, the **cranium** (which means "helmet" in Greek) and the face. The cranium houses the brain and balancing mechanisms and also provides openings for the ears and a structure for the joint of the jaw; the face provides a structure for the mouth and jaw, and includes muscles of expression, sockets for the eyes, and the nasal cavities.

The bones that make up the skull are quite complex; for purposes of study, it is best to consider the skull by looking at an actual model, since the bones of the skull are three-dimensional, varied in shape, and they defy description.

The cranium is made up of five bones: the **parietal,** the **temporal,** the **sphenoid,** the **ethmoid,** and the **occipital bone.** The **parietal** bones (*paries,* meaning "a wall") form the sides and top of the skull. The **frontal** bone (*frons,* the forehead) forms not only the forehead, but also the roof of the eyes and nasal passages. The **sphenoid** bone (meaning "a wedge") forms the front of the base of the skull and joins the other bones of the cranium. The **ethmoid** (which is a Greek word meaning "sieve") forms part of the base of the cranium and the nasal cavities. The **temporal bones** (*tempus,* time), more commonly known as the temples, form the sides and the rest of the base of the skull. The **occipital** bone (*ob, caput,* "against the head") forms the back and base of the skull (Fig. 3).

The face is made up of the **nasal** bones; the **turbinate** (*turbo,* a whirl); **vomer** (*vomer,* a ploughshare); the **lachrymal** bones (*lachryma,* a tear); the **zygomatic,** or cheek bones (this bone is sometimes called the **malar,** which also means "cheek"); the bones of the **palate;** the **superior maxilla,** or upper jaw ("maxilla" means "jaw-bone"), and the **inferior maxilla,** or lower jaw, also called the **mandible** (Fig. 3).

All the bones of the skull are connected by **sutures,** or seams, which are uneven surfaces that fit together and are held firmly in place by fibrous tissue.

Fig. 3.   The skull

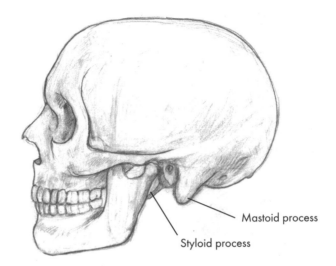

Mastoid process

Styloid process

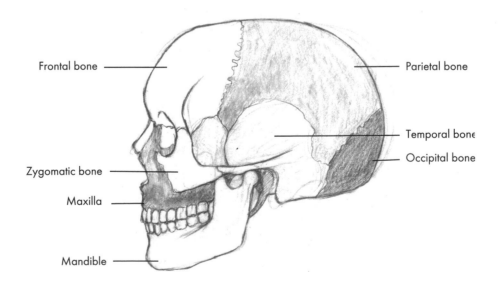

Frontal bone

Parietal bone

Temporal bone

Occipital bone

Zygomatic bone

Maxilla

Mandible

Fig. 4.   The base of the skull

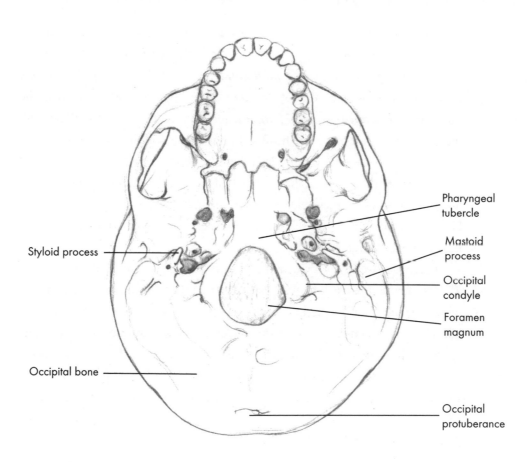

Pharyngeal tubercle

Mastoid process

Styloid process

Occipital condyle

Foramen magnum

Occipital bone

Occipital protuberance

At birth, the bones of the skull are not fully formed, leaving gaps that can be felt between the sutures; after a time, the sutures become fully formed and the bones join firmly together.

Let's now look at the base of the skull (Fig. 4), which as mentioned above is formed mostly by the occipital and temporal bones. The most prominent feature on the base of the skull is the *foramen magnum,* the large hole through which the brain stem passes. On either side of the foramen magnum are two rounded bumps called the *occipital condyles.* These bumps fit like rockers into two depressions on the top vertebra of the spine, the *atlas,* to form the head/neck or *atlanto-occipital joint,* the joint where the skull articulates with the spine (you can move the head at this joint by nodding the head up and down). Toward the back of the skull there is a bump on the occiput where it rounds to form the back of the head called the *occipital protuberance*—this area sometimes becomes tender when we lie in the semi-supine position. This is an important point because the neck muscles—the superficial ones that pull the head back—attach here.

There are two other important points at the base of the skull (Figs. 3 and 4). Just behind the ear lobe you can feel a bump called the *mastoid process,* which means "teat" or "nipple." This, as we'll see in a moment, is an important point of attachment for the *sternocleidomastoid* muscle, the large muscle of the neck that attaches to the sternum. Another crucial attachment point for muscles of the neck and throat is the *styloid process* (*stylos,* a pillar), which is a sharp projection near the mastoid bone (it is set in from the sides of the skull and so is not visible or palpable). The mastoid and styloid processes are actually part of the temporal bone and are closely linked with the point at which the skull balances on the spine. The temporal bones also house the ear canals and balancing mechanisms of the inner ear, which are located at this balancing point of the skull near the styloid process. Next we will look in more detail at the atlanto-occipital joint and some of the key muscular attachments to the base of the skull.

4. Base of the Skull and Its Attachments

In the last section we looked at the bones of the skull. Let's look now at the muscles attaching to the base of the skull. Most people—even those who are familiar with anatomy—think of the body as an aggregate of many separate parts. The biceps muscle of the arm, the hamstring muscles of the leg, the back muscles—all are viewed piecemeal, and without any overall organizing elements. This is particularly true of the head/neck region, which, viewed as a whole, seems to be made up of a network of complex and indecipherable muscles that must be learned one by one. However, when we consider that the various muscles in this region relate directly or indirectly to the base of the head and its balance on top of the spine, it becomes possible to sort the muscles in this region into distinct and understandable systems. So let's start by getting an overview of the different muscle systems of the head/neck region as they relate to the base of the skull.

To begin with, the base of the skull corresponds not with the underside of the jaw, but the cheekbone. This is very important, because when we confuse the head with the jaw and parts of the neck, we lack a clear picture of what the head is and therefore cannot accurately "direct" the head in movement. It also means that we have a fuzzy picture of what the neck is; functionally, the neck goes much higher than we think, and the head articulates with the spine not at the level of the jaw but at the level of the cheekbone, at the point just between the ears.

The base of the skull has a number of crucial points on it that we must be clear about before going on. Obviously the main purpose of the large round mass of the skull is to house the brain. Right in the middle of the underside of the skull is the large hole called the ***foramen magnum***. The brain stem passes down through this opening to form the spinal cord, which runs down

Fig. 5. Flexors and extensors attaching to base of skull

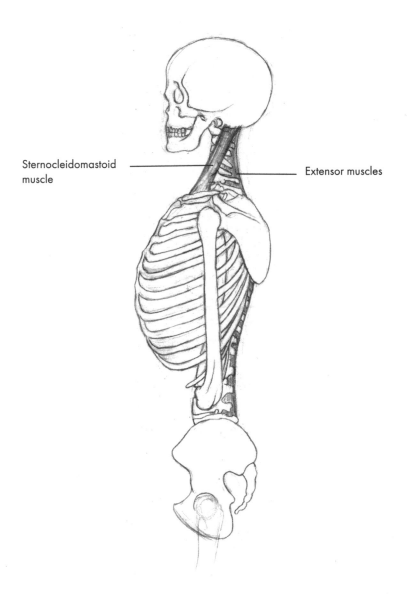

Sternocleidomastoid muscle

Extensor muscles

the length of the spine, sending nerves into every part of the body. (There are actually several "foramen," or openings, in the skull; "foramen magnum" means "great opening.") Just in front and to the sides of the foramen magnum on the underside of the skull are the ***occipital condyles,*** two smooth bumps where the head articulates on the first vertebra of the spine to form the ***atlanto-occipital joint*** (meaning the joint formed by the atlas, or first vertebra of the spine, and the occiput; see Fig. 4). This basic point of articulation of the skull with the spine is common to all vertebrates; in humans, of course, the spine is vertical, and the head is balanced on top of the spine here. The main movement we can perform at this point is head nodding, or flexion and extension of the skull.

However, the head is not balanced evenly at this point. If you look at the location of the occipital condyles, you'll see that the center of gravity of the head is forward of the point of balance, causing the head naturally to nod or fall forward. This tendency of the head to fall foward counteracts the pull of muscles at the back of the neck, naturally keeping these muscles on stretch. This facilitates lengthening of the spine and stimulates the reflexes that support the body against gravity. So this uneven balance of the skull on the spine is an important detail about the anatomy of the skull to keep in mind.

This point where the head is poised on the top vertebra of the spine is the ***frontal plane,*** also called the ***coronal plane of balance***—the plane that bisects the front and back halves of the body (see Fig. 1). The semi-circular canals, which register movement of the head in space, are located in this plane, three on either side of the head about where the ears are but in from the sides, each one registering orientation in one of the three spatial dimenstions—the vertical plane, the horizontal plane, and the transverse plane. So we can see that this point where the head balances on the spine is a very crucial point in our anatomy.

The first system of muscles that relates to the base of the skull and its balance on the spine is composed of the deep postural muscles that support us

Fig. 6.   Muscles supporting hyoid bone and larynx

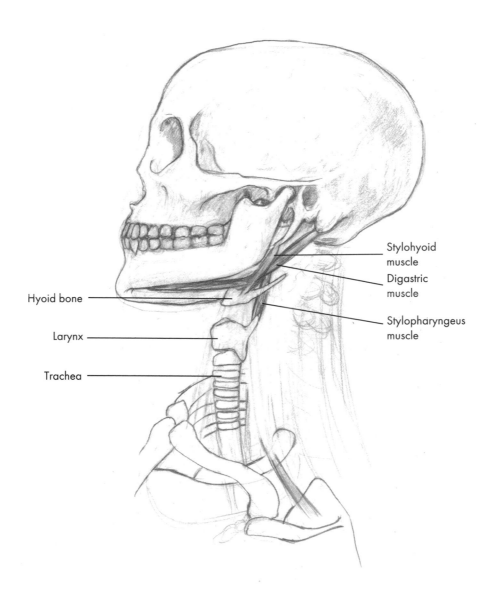

Stylohyoid muscle

Digastric muscle

Stylopharyngeus muscle

Hyoid bone

Larynx

Trachea

against gravity. These muscles run along the spine and right up to the skull around the foramen magnum. We saw in the last section that the back and underside of the skull are formed mainly by the *occiput.* The bump on the back of the occiput, the *occipital protuberance,* is the main point of attachment for the neck muscles that pull the head back. These muscles form the beginning of the system of extensor muscles that run down the length of the back—the main mass of muscles that support us against gravity (Fig. 5). You can easily identify this point on the skull because the bump is fairly prominent in most people.

On either side of the back of the skull, just behind the ear lobe, is the *mastoid process*—the bony bump to which the large muscle on the front of the neck, the *sternocleidomastoid,* attaches. This muscle, which as its name suggests connects the sternum and clavicle with the mastoid process, forms an important part of the flexor system of muscles on the front of the body, and relates the length or support of the front of the body to head balance (Fig. 5). There are also smaller and deeper muscles that directly connect the skull to the spine, just where the skull rests on the atlas; these muscles are part of the system of small muscles that run the length of the spine, and they are crucial to posture. So the flexor mass of muscles in the front, the extensors in the back, and the deeper postural muscles all attach to the base of the skull and so are directly related to the balance of the head on the spine.

Another system of muscles that relates to the base of the skull are the muscles of the vocal mechanism. Just in front of the mastoid process is the *styloid process* (*stylos,* a pillar), the small spikes of bone lying on either side of the occipital condyles. Like the mastoid processes, these bones also lie in the coronal plane, the point at which the head is balanced on the top vertebra of the spine. (The semi-circular canals, which register movement of the head in space, are embedded in the skull near the styloid processes.) Just below the line of the jaw on the neck is a little bone, the *hyoid* or tongue bone, as it is sometimes called—a horseshoe or U-shaped bone that lies just above the

Fig. 7.   Base of the skull and muscles of the throat

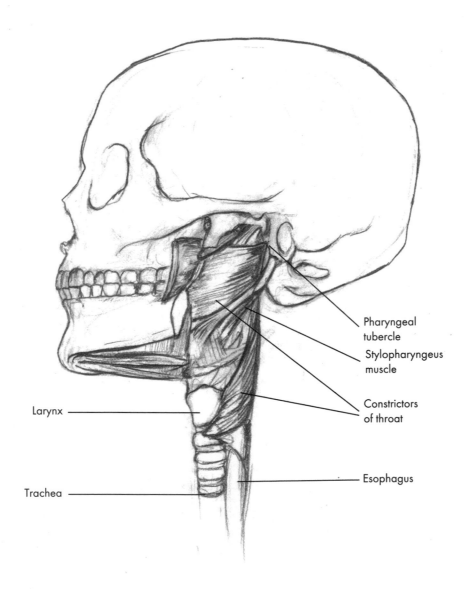

Pharyngeal tubercle

Stylopharyngeus muscle

Constrictors of throat

Larynx

Esophagus

Trachea

larynx and windpipe and which forms the root of the tongue. (Its name comes from the Greek letter *"upsilon,"* which is shaped like a U.) You can feel it if you pinch the throat just above the larynx and under the line of the jaw, and particularly if you swallow or wag your tongue. The larynx, the mechanism concerned with producing sound, is located just below the hyoid bone. It has an outer protective structure called the thyroid cartilage, which is directly hung, or suspended, from both the hyoid bone (by the *thyrohyoid* muscle) and from the styloid process (by the muscle *stylopharyngeus*). The hyoid bone, in turn, is suspended directly from the styloid process by the *stylohyoid* muscle, linking the tongue and the larynx to the base of the head. The tongue is also directly connected to the styloid process by the muscle *styloglossus,* forming another crucial link between the tongue and head balance. Another muscle connecting the larynx to the head is the *digastric muscle,* which runs from the mastoid process, through a loop on the hyoid bone, and on to the jaw, forming another part of the network of muscles that support the throat and vocal mechanisms from the base of the skull (Fig. 6).

There are several other muscles that form the network of the larynx (we'll cover these later in more detail), but the muscles mentioned above are some of the principal ones that demonstrate the direct connection of the larynx and vocal structures to the head. The point to keep in mind is that the vocal mechanism is suspended from the skull and in particular from the styloid process, which is why good vocal use is so dependent on head balance. If the head is pulled back by tensing the muscles of the neck and depressing the larynx while vocalizing, this general pattern of misuse (which also involves the trunk and legs) ultimately interferes with proper vocal function. It is the structures connected with depressing the larynx and pulling back the head that we'll be primarily concerned with in the next several sections on the head/neck region.

The pharynx, or throat, is another structure directly connected with the base of the skull and its balance on the spine. Just in front of the foramen magnum is a bump called the *pharyngeal tubercle* (see Fig. 4). The muscles

Fig. 8.   Muscles and joint of jaw

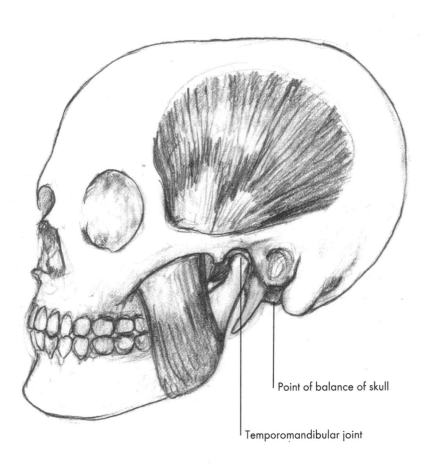

Point of balance of skull

Temporomandibular joint

forming the back wall of the throat, the constrictors of the throat, originate here, so that the tube of the throat is essentially suspended from this point on the skull, contributing to the tendency to pull down in front. The sides of the wall of the pharynx are also directly connected to the styloid process by the muscle *stylopharyngeus* and so form another crucial link between the structures of the throat and head balance (Fig. 7).

There is one other structure directly related to the balance of the head that is important to keep in mind—namely the jaw or mandible. The jaw articulates with the skull at a point very close to the coronal plane of balance, and it is also suspended from muscles attaching to the temporal bone of the skull (Fig. 8). In addition, several muscles on the underside of the jaw attach to the hyoid bone (which as we saw is also hung from the skull), and one connects directly to the mastoid process. So the jaw is directly hung from the skull and intimately affects and is affected by head balance.

The head, then, is balanced on the top vertebra of the spine, forming the atlanto-occipital joint; movements of the skull are detected by the vestibular mechanisms of orientation and balance, which are located on the coronal plane of balance; this point of balance is also crucial to the muscular balance of the entire flexor and extensor system of muscles that support us against gravity. The throat is hung from the base of the skull; the larynx and vocal mechanisms are also suspended from the skull and intimately connected with head balance (a crucial connection from the point of view of movement educators who work with the voice and breathing); the jaw is slung from the skull; and the tongue directly connects with the base of the skull. In short, the flexors in the front of the body, the extensors in back, our inner and middle ears, our innards, our jaw, the tongue, and the vocal mechanism are all suspended from or located at the base of the skull and directly or indirectly relate to the balance of the head on the spine. As we go along, we'll look at these various structures in more detail.

35

5. Muscles of the Face and Jaw

Let's turn now to the muscles of facial expression and the jaw (Figs. 9 and 10). The forehead muscle, *frontalis,* is actually part of a larger sheet of muscle and connective tissue that runs from the forehead right over the scalp to the occiput in back, called the *epicranius muscle.* The forehead and occipital portions are muscle, connected by a tendinous sheet in between.

Another muscle, the appropriately named *corrugator,* lies in between the eyebrows; it is the frowning muscle and causes the vertical wrinkles in the forehead. *Procerus* lies just below corrugator and also relates to frowning.

There are some tiny muscles just beneath the ear that wiggle the ears; they correspond to the more developed muscles in dogs that make it possible for them to adjust the ears to pick up sounds.

Surrounding the orbits (the eye sockets) is a sphincter muscle, *orbicularis oculi.* Its function is to narrow and protect the area around the eyes. There is another muscle that elevates the eyelid, as well as six muscles within the orbit that turn the eyeball.

There are several muscles in the nasal region—the *dilators, compressors,* and *depressors* of the nostrils, as well as muscles that wrinkle the nose. These muscles, as their names suggest, dilate, compress, and depress the nostrils; the actions of these muscles can be discovered by experimentation and are actually quite important to breathing and vocalization.

There are several muscles around the cheek and the area above the upper lip that relate to facial expressions such as sadness and laughing. *Levator labii superioris* and *levator anguli oris* raise the lip and mouth. *Zygomaticus major* and *zygomaticus minor* arise from the cheekbone and pass down to the mouth; these muscles draw the mouth back and up and back and down, as in

Fig. 9.   Muscles of facial expression

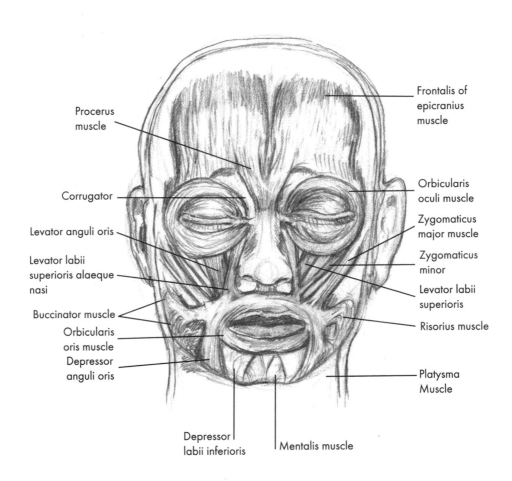

Procerus
muscle

Frontalis of
epicranius
muscle

Corrugator

Orbicularis
oculi muscle

Zygomaticus
major muscle

Levator anguli oris

Zygomaticus
minor

Levator labii
superioris alaeque
nasi

Levator labii
superioris

Buccinator muscle

Risorius muscle

Orbicularis
oris muscle

Depressor
anguli oris

Platysma
Muscle

Depressor
labii inferioris

Mentalis muscle

laughing and expressing sadness. In conjunction with the muscles around the eyes and nose, these are the smiling muscles needed to brighten the face and "place" the voice forward into the face, and they are crucial to vocalization and opening the throat. This entire area, including the nasal region and palate, can be mobilized and widened as one system, releasing the entire musculature of the face in front right around to the joint of the jaw.

Orbicularis oris is a circular muscle (not technically a sphincter muscle) that controls movements of the lips. Below the lips, ***mentalis, depressor anguli oris,*** and ***depressor labii inferioris*** act on the lips, the mouth, and the chin.

Risorius and *buccinator,* which means "trumpet" in Latin (from the word *"buccina"*), form the wall of the cheek. Risorius retracts the corners of the mouth; buccinator compresses the cheeks in order to keep food between the teeth; it also helps expel air when the cheeks are distended as in trumpet playing—hence the name!

Platysma is a large, thin sheet of muscle covering the neck and lower jaw. It arises from the fascia of the muscles of the shoulder and chest, passes over the neck, and attaches to the lower part of the face. It acts on the jaw and lips and aids in the expression of sadness.

Now to the muscles of the jaw (Fig. 10). The jaw bone, called the *mandible,* articulates with the temporal bone of the skull. The jaw serves a very basic function—namely, to seize and chew food. It is supplied by the 5th cranial, or trigeminal nerve, which is one of the primary nerves of the face and jaw and is responsible for the primitive actions of seizing and chewing food. It is a branch of this nerve that the dentist anesthetizes when you are having a filling done.

The joint where the jaw articulates with the skull is called the ***temporomandibular joint*** (the joint formed by the temporal bone and the mandible; see Fig. 8). The temporomandibular joint isn't purely a hinge-like joint, as

Fig. 10. Muscles of the jaw

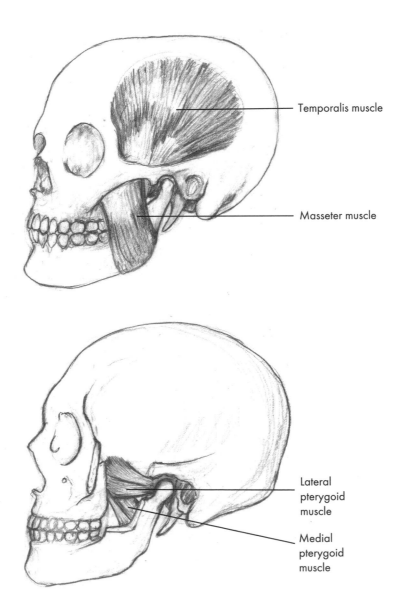

Temporalis muscle

Masseter muscle

Lateral pterygoid muscle

Medial pterygoid muscle

many people think. It is both hinged and slung from the skull, making it possible not only to snap the jaw shut, but to slide the lower jaw forward and back and to move it sideways.

"TMJ syndrome" refers to the painful condition of tension and dysfunction associated with the temporomandibular joint; rather severe problems can develop when the jaw, whose correct function is dependent on the postural system and whose muscles are so powerful, is subjected to muscular stresses that may impinge on a nerve.

The jaw basically does three things: it snaps shut, it clamps down, and it grinds. The main muscle responsible for snapping is the ***temporalis,*** which is a broad, powerful muscle that originates broadly at the temporal region on the side of the head and converges to insert into the coronoid process of the jaw.

The ***masseter*** muscle originates at the cheekbone and inserts into the ramus of the jaw. Its function is to raise the lower jaw and to clamp it shut.

The ***lateral*** and ***medial pterygoid*** muscles arise from the cheekbone and palate areas of the skull and insert into the condyles and ramus (the back parts) of the jaw bone. Their function, particularly when acting alternately, is to grind the jaw. All three of the jaw muscles acting together produce the movements of chewing and grinding food.

6. Suspensory Muscles of the Larynx

We saw earlier that the throat and larynx are suspended from the skull, in particular from the styloid process. There are several other muscles that anchor the larynx—particularly from below. Let's now look in more detail at the larynx and the muscular scaffolding, or "suspensory mechanism" as it is sometimes called, that supports it.

To review the muscles connecting the larynx and pharynx with the skull: *stylohyoid* connects the hyoid bone to the styloid process. The muscle *digastricus,* which means "having two bellies," originates at the mastoid process, runs through a loop on the hyoid bone, and continues on to insert into the jaw. *Stylopharyngeus* directly connects the thyroid as well as the sides of the pharynx with the styloid process.

I want to look now at the larynx itself—the vocal mechanism—and the entire network of muscles that support it (Figs. 11 and 12). As we've seen, the larynx, whose main outer structure is called the thyroid, or shield, cartilage, lies below the hyoid bone. A number of muscles directly support the larynx by attaching to the thyroid cartilage (Fig. 11). First, the thyroid is connected just above to the hyoid bone; this pair of muscles is called the *thyrohyoid.* Below, the thyroid is connected to the sternum by another pair of muscles— this is the *sternothyroid* muscle.

Two other muscles help form the scaffolding for the larynx. *Stylopharyngeus,* which we've already mentioned, directly connects the thyroid as well as the sides of the pharynx with the styloid process. And *cricopharyngeus* anchors the thyroid cartilage of the larynx directly back to the esophagus.

So the larynx itself is suspended from the hyoid bone, which has a muscular connection to the styloid process via stylohyoid and to the mastoid

Fig. 11. Suspensory muscles of the larynx

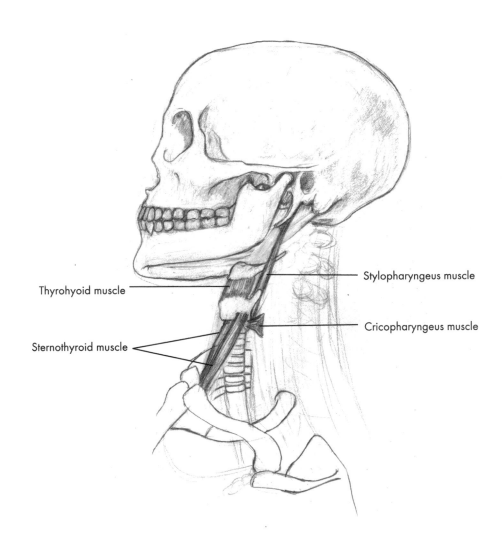

Thyrohyoid muscle

Sternothyroid muscle

Stylopharyngeus muscle

Cricopharyngeus muscle

Fig. 12. Suspensory muscles of the larynx (cont.)

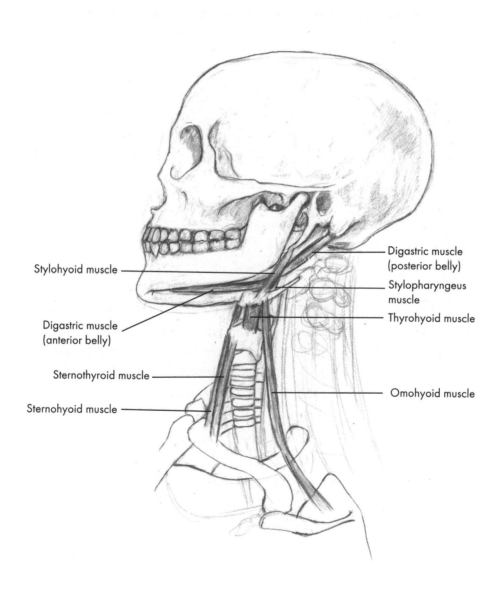

Stylohyoid muscle

Digastric muscle (anterior belly)

Sternothyroid muscle

Sternohyoid muscle

Digastric muscle (posterior belly)

Stylopharyngeus muscle

Thyrohyoid muscle

Omohyoid muscle

process via digastricus; it is more directly connected back and up to the skull via stylopharyngeus; it is connected below to the sternum via sternothyroid; and it is connected behind to the esophagus via cricopharyngeus—a complex web of muscular support that extends up, down, back, and up and back from the larynx.

Based on the directions indicated, we can infer what actions are performed by these muscles. The thyrohyoid and stylopharyngeus, as well as the muscles of the palate (which we'll talk about later), can be seen to pull the larynx up and back; they are therefore elevators of the larynx. The sternothyroid and cricopharyngeus muscles pull the larynx down and back, and are therefore depressors. All these muscles are actively involved in antagonistically supporting the larynx within a muscular scaffolding when singing.

There are several other muscles that indirectly form the scaffolding for the larynx (Fig. 12). ***Omohyoid,*** which actually originates at the scapula, runs through a sheath of muscle in the neck, and from there changes angle to run directly up to the hyoid bone. Like digastricus, it has two bellies and runs in an almost sling-like arrangement to provide an indirect part of the muscular scaffolding for the hyoid bone and larynx. There is also a muscle, ***sternohyoid,*** that makes a direct connection of the hyoid bone with the sternum.

To complete the picture, there are two muscles on the underside of the jaw, ***mylohyoid*** and ***geniohyoid,*** that form the floor of the jaw and relate it to the hyoid bone (Fig. 14). Mylohyoid is a fan-shaped muscle whose fibers run at a downward angle from the sides of the jaw to join with the hyoid bone, covering the base of the jaw. Geniohyoid, which lies above mylohyoid, runs from the front of the jaw to the hyoid bone.

Muscles that pull in an upward direction on the hyoid bone, such as the mylohyoid and digastricus, are false elevators of the larynx; the tendency to overuse these muscles can often be detected in popular singers, who appear to have a kind of double-chin from overworking these muscles.

7. The Tongue

The tongue, or lingual region, is often neglected by those interested in movement and even vocalization. Often when we speak of the tongue we thinking only of the *body* of the tongue; in fact, there are a number of extrinsic muscles attaching to the tongue that control its movement. All these muscles—the muscles attaching to the tongue and the tongue muscle itself—relate directly to the structures of the throat, the larynx, and head balance.

The intrinsic muscle of the tongue (the body of the tongue itself) is made up of fibers of muscle running in various directions. It lies at the floor of the mouth and extends back into the pharynx, where it connects at its root to the **hyoid bone,** sometimes called the "tongue bone," which is the little U-shaped bone above the larynx (which, as we saw, gets its name from its resemblance to the Greek letter *upsilon*). You can feel the connection of the tongue with the hyoid bone if you grasp the hyoid bone and larynx, which is directly connected to the hyoid bone, between your thumb and forefinger and then wag your tongue: you can feel the hyoid bone and the larynx move back and forth as the tongue moves.

There are four extrinsic muscles that join with the fibers of the body of the tongue, and whose function is to support and move the tongue (Fig. 13). We've already seen one of the extrinsic muscles of the tongue that relates the tongue to the head and its point of balance on the spine—the *styloglossus muscle,* which originates at the styloid process and joins into both sides of the body of the tongue. *Hyoglossus* attaches from the hyoid bone into the sides of the tongue. The third extrinsic muscle of the tongue is *genioglossus,* which runs vertically from the jaw at the inside of the chin up to the mid-line of the tongue. The fourth and final extrinsic muscle of the tongue is the *palatoglossus* muscle, which compresses the tongue and soft palate when

Fig. 13. The tongue

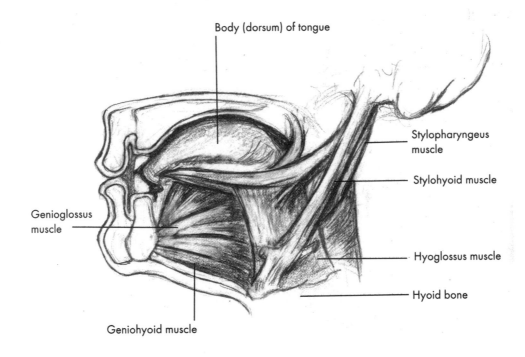

Body (dorsum) of tongue

Stylopharyngeus muscle

Stylohyoid muscle

Genioglossus muscle

Hyoglossus muscle

Hyoid bone

Geniohyoid muscle

Fig. 14. Muscles on the floor of mouth

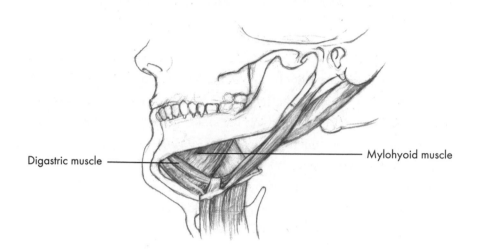

Digastric muscle ———— Mylohyoid muscle

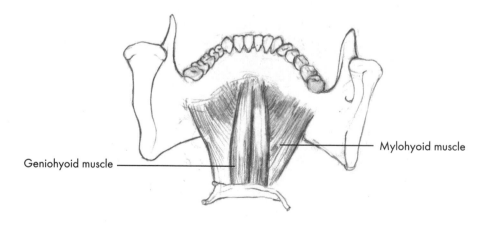

Geniohyoid muscle ———— Mylohyoid muscle

swallowing. We'll look further at this muscle when we discuss the muscles of the palate. "*Glossus*," which is a Greek word, also means "tongue."

The extrinsic muscles of the tongue make it possible to move the tongue in virtually all directions. The fibers of the tongue itself make it possible to move its tip and perform other complex movements involved not only in speech, but also in positioning food for chewing and swallowing.

Having looked now at the tongue and the muscles attaching to the hyoid bone, we can also get a fuller picture of the muscles that form the floor of the mouth (Fig. 14). There is the large fan-shaped muscle at the floor of the mouth, ***mylohyoid,*** attaching along the sides of the jaw to the hyoid bone. Above this are the lengthwise fibers of ***geniohyoid*** running from the front of the chin to the hyoid bone. We saw earlier that the ***digastric*** muscle originates from the mastoid process, runs through a sling on the hyoid bone, and then passes on to the front of the jaw, so that the front, or anterior, belly of digastricus forms part of the floor of the mouth.

8. Muscles of the Palate

The palate, or roof of the mouth, is another crucial area of the musculature of the head/neck region that is often neglected, but it is in fact crucial to healthy breathing and vocal use (Fig. 15). The palate consists of two regions: the hard palate, which forms the roof of the mouth; and the soft palate, which lies behind the hard palate and is made up of soft tissue not unlike that of the tongue, forming an arched structure at the back of the mouth. In the illustration of the throat (Fig. 16), you can see the opening of the mouth, with the hard palate in front and the soft palate behind. The illustration also shows the different levels of the pharynx, corresponding to the nasal cavity, the oral cavity, and the larynx.

The *uvula,* which is the visible structure hanging down from the soft palate at the back of the mouth, is actually controlled by a muscle, the *uvular muscle,* which raises it and moves it back (Fig. 15).

Just in front of the uvula are two arches or pillars; if you look into someone's throat, you can see these arches at the back. You can also see the tonsils, which lie along them. These pillars are formed by two muscles, *palatoglossus* and *palatopharyngeus.* Palatoglossus forms the anterior (front) pillar of the soft palate, passing from the tissue at the front of the soft palate and sloping down on both sides to join into the sides of the tongue.

Palatopharyngeus forms the posterior (back) pillar of the soft palate, dropping down from the soft palate to join the stylopharyngeus muscle and thyroid cartilage. These two muscles depress the palate during eating and swallowing. They also help to elevate the larynx during eating and vocalization.

Above the two depressors of the palate are the muscles that elevate the palate. *Levator veli palatini* attaches from the skull and passes down into the

Fig. 15. Muscles of palate

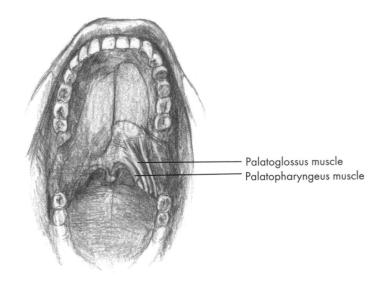

Palatoglossus muscle
Palatopharyngeus muscle

(Anterior view of soft palate)

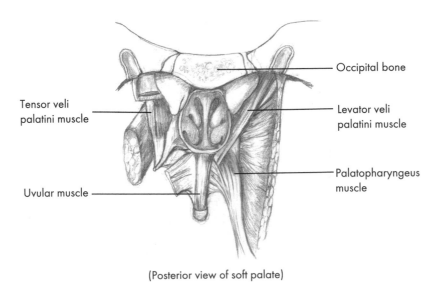

Occipital bone

Tensor veli palatini muscle

Levator veli palatini muscle

Uvular muscle

Palatopharyngeus muscle

(Posterior view of soft palate)

fibers of muscle that form the upper part of the arches, or pillars, of the palatopharyngeus and palatoglossus muscles. When we yawn or deepen the voice, which raises the palate, it is this muscle which comes into play—a crucial one for voice-users.

Tensor veli palatini attaches from the sphenoid bone of the skull and runs vertically down and sideways to assist in tensing and elevating the soft palate.

The *salpingopharyngeus muscle* (not shown in illustration) passes from the skull near the inner ear to join with the palatopharyngeus and also assists in raising the palate. These elevators of the palate lie above the soft palate and are not visible to the eye.

So the two "pillar" muscles, palatopharyngeus and palatoglossus, depress the palate; the two "palatini" muscles, as well as salpingopharyngeus, elevate it; and the uvula muscle, azygos uvulae, raises the uvula at the very back of the palate.

The structure of the palate can be likened to a diaphragm that is capable of being raised or depressed. We raise and tense the palate during swallowing, and we raise it during vocalization and normal respiration. By raising the back of the tongue against the palate, we can also seal off the oral passage, making it possible to keep food and air passages separate during swallowing. It should also be possible, by leaving the back of the tongue raised, to breathe exclusively through the nostrils while opening the mouth or moving the lips and tongue. However, most people habitually depress the palate, collapse the throat and tongue, and pull back the head while sleeping, breathing, and vocalizing, which leads to mouth-breathing and snoring, tends to block the nasal passages, and interferes with the mobility and openness of the throat. Skilled voice-users, instead of collapsing the palate during vocalization, can raise it at will; the result is a much healthier, toned, and open throat. You can observe the raising of the palate by looking down someone's throat or using a mirror to observe your own while yawning: the pillars of the soft palate will visibly

53

Fig. 16. Muscles of the throat

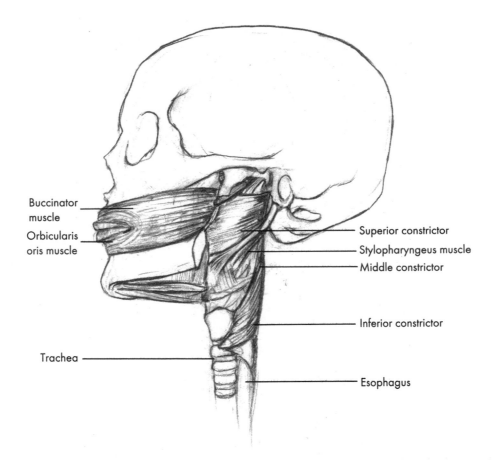

open and move up. Learning to alter the shape of the pharynx by raising the palate has a marked effect on resonance and vocal quality; it is also one of the goals of performing such vocal procedures as the "whispered 'ah'" or humming, which both help to open the throat.

9. Muscles of the Throat

Let's turn now to the throat. From a phylogenetic point of view, the throat is much older than the musculature of the larynx, since it relates to acquiring and processing food, which evolved long before the ability to vocalize and even to breathe. When we look at the musculature of the larynx, it appears to be a complex web; looked at from the point of view of eating and digestion, the muscles of the throat form a tube whose opening is of course the mouth, for taking in food and passing it down the esophagus. The entire digestive tract is lined with muscle, but in musculoskeletal anatomy we are concerned only with the throat.

As we saw in the chapter on the muscles attaching to the base of the skull, the muscles of the throat, or *pharynx,* are suspended from the *pharyngeal tubercle* on the base of the skull (see Fig. 7). Three muscles form the swallowing muscles of the throat: *superior constrictor, middle constrictor,* and *inferior constrictor* (Fig. 16). When food or liquid is taken into the pharynx, these muscles are automatically brought into play, alternately contracting and squeezing food downward into the esophagus and stomach.

The passageway of the throat is entered by the mouth, which is partly formed by *orbicularis oris,* the muscle around the mouth, and *buccinator,* which forms the walls of the cheek (Fig. 16). The palate muscles, which depress during swallowing, form the upper part of the back of the throat and help to draw food downwards. The three constrictor muscles form the back wall of the pharynx; the front wall is formed by the back of the tongue. At this point, at the level of the larynx, the throat divides into the *trachea* or windpipe and the *esophagus,* the trachea being in front and the esophagus in back (Fig. 17). When food enters the mouth and we prepare to swallow,

Fig. 17. The pharynx

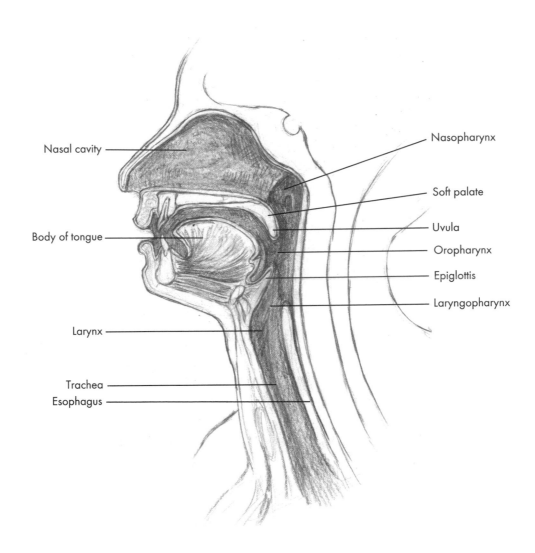

Nasal cavity

Body of tongue

Larynx

Trachea
Esophagus

Nasopharynx

Soft palate

Uvula

Oropharynx

Epiglottis

Laryngopharynx

the ***epiglottis,*** a large flap of tissue just above the opening of the larynx, automatically closes down so that the opening of the larynx and windpipe is closed off; food is then passed down into the esophagus and stomach.

Several seemingly disparate structures, then, are involved in eating and swallowing, forming a continuous system when looked at from this point of view. The mouth, jaw, tongue, and cheeks are involved in chewing. The tongue is elevated to drive the food back into the throat. The palate muscles raise the palate to allow the food through and then constrict in order to drive the food downwards. And the constrictors automatically begin forcing food into the esophagus. From this point of view, the tongue and palate muscles are part of the digestive system; even the muscles of the larynx, which we usually think of in relation to vocalization, elevate the larynx to allow food to come in and then depress during swallowing. So virtually all the muscles of the head and neck, including the muscles connecting to the larynx and hyoid bone, are involved in eating and swallowing and function as part of this system. They can be misused in this as in any other activity.

10. The Larynx

Let's look now at the intrinsic muscles of the larynx—the vocal folds and the muscles that control them. In the discussion on the suspensory muscles of the larynx, we saw that the larynx is supported within a scaffolding or web of muscles. These muscles act upon the larynx but are not part of the larynx itself, and so are extrinsic muscles of the larynx. The muscles of the larynx itself are the intrinsic muscles.

The muscles of the larynx are highly specialized and somewhat complex, but when we consider the actual function of the larynx, it is easier to understand its anatomy. When we exhale normally, air passes unimpeded through the windpipe and out the mouth or nose. When we think of a sound, the larynx, which is located above the windpipe and through which the air passes, draws together the two vocal folds, which begin to vibrate as air forces its way between them, creating sound waves that resonate to create the fully formed sound of the human voice.

The larynx, then, is basically a vibration mechanism. It contains the oscillators that make sound (the vocal folds), and it can bring them together so that they will vibrate when air (which is the power source) passes between them, and draw them back apart during normal breathing. It can also tense and stretch the vocal folds in various ways in order to alter the volume, pitch, and types of vibration that occur. In short, the muscles of the larynx control the vocal folds, which are themselves muscles.

The larynx itself is comprised of three cartilages which form the housing for the vocal folds—the *thyroid* or shield cartilage; the *cricoid* or ring cartilage; and the *arytenoid* or *pyramid* cartilages (Fig. 18). The thyroid cartilage is the main structure protecting the vocal folds; it forms the prominence on some people's throats that we call the Adam's apple. (It is also the

Fig. 18. The larynx

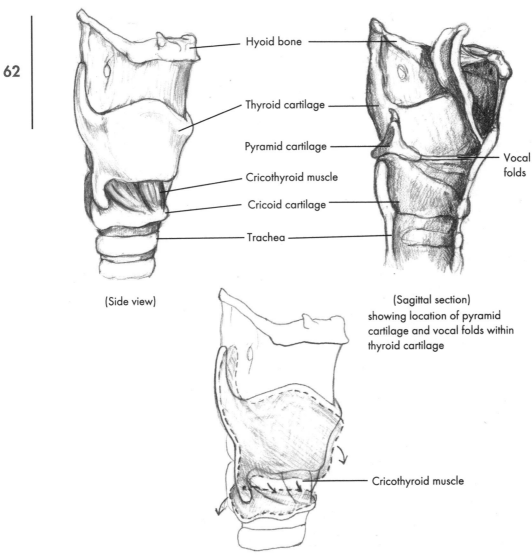

Hyoid bone

Thyroid cartilage

Pyramid cartilage

Cricothyroid muscle

Cricoid cartilage

Trachea

Vocal folds

(Side view)

(Sagittal section)
showing location of pyramid
cartilage and vocal folds within
thyroid cartilage

Cricothyroid muscle

Action of cricothyroid muscle: stretches vocal cords

main structure to which the suspensory muscles attach, supporting the larynx within a muscular network). Below the thyroid cartilage is the cricoid cartilage. And sitting on the back part of the ring cartilage within the thyroid cartilage are the two pyramid cartilages, which are capable of rotating in position on the ring cartilage.

The vocal folds—the muscles that come together and vibrate when the air hits them to create sound—are suspended within the thyroid cartilage between the two pyramidal cartilages and the inner, front wall of the thyroid cartilage (Figs. 18 and 19). When the pyramid cartilages rotate, this brings the vocal folds together so that they vibrate as the air passing out of the lungs forces them open; when the pyramid cartilages rotate back, the vocal folds are separated to allow air to pass through freely again (Fig. 19).

Three sets of muscles rotate the pyramid cartilages: the ***posterior cricoarytenoid*** muscles, which open, or abduct, the vocal folds; and the ***transverse arytenoid*** and ***lateral cricoarytenoid*** muscles, which close, or adduct, the vocal folds (Fig. 19). For convenience these latter muscles can also be called the "closers," since they close the space between the vocal folds. Notice that there are two, not one, set of adductors. This is because the pyramid cartilages are able to close the vocal folds in two different ways. First, the pyramid cartilages can rotate in such a way that the two vocal folds lie against each other along their entire length, so that the air passing out of the trachea causes the vocal folds to oscillate, creating a normal vocalized sound. Second, the pyramid cartilages can rotate in such a way that although the vocal folds are approximated, there is a chink of space left between them. Air coming out of the trachea is then able to escape through this chink, causing the "whispered" sound of rushing air. So there are two ways of closing, or adducting, the vocal folds, and that's why there are two sets of adductors.

Knowing that there are two ways of adducting the vocal folds is useful for teachers who work with the voice, for two reasons. First, it provides us with a means of engaging the vocal folds in a way that is not associated with our

Fig. 19. Intrinsic muscles of the larynx

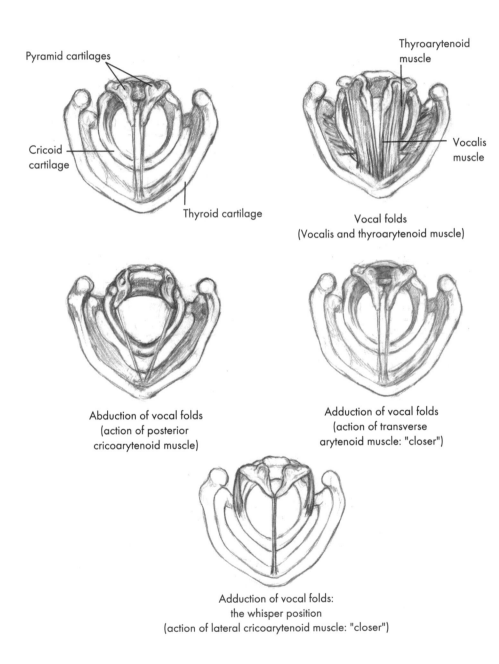

Pyramid cartilages

Cricoid cartilage

Thyroid cartilage

Thyroarytenoid muscle

Vocalis muscle

Vocal folds
(Vocalis and thyroarytenoid muscle)

Abduction of vocal folds
(action of posterior
cricoarytenoid muscle)

Adduction of vocal folds
(action of transverse
arytenoid muscle: "closer")

Adduction of vocal folds:
the whisper position
(action of lateral cricoarytenoid muscle: "closer")

usual vocal habits. Our harmful habits of vocal use are associated with normal closure of the vocal folds, so that by whispering instead of vocalizing normally, we do not bring these harmful habits into play. Whispering thus provides a ready means of learning to vocalize without the usual pattern of harmful use. Second, it explains the significance of why the pyramid cartilages work the way they do. One way of rotating the pyramid cartilages fully closes the vocal folds; another only partially closes the vocal folds and corresponds to whispering.

In addition to being able to close the space between the vocal folds (also known as the ***glottis***), the larynx is capable of stretching the vocal folds. This is necessary because when air passes between the closed, or approximated, vocal folds, they must also be somewhat taut in order to vibrate properly. This action is performed by the ***cricothyroid*** muscle, which attaches to the cricoid cartilage and the thyroid cartilage, and which pulls the two cartilages together in such a way that the distance between the two extremities of the vocal folds is increased, causing them to stretch (Fig. 18). Because of the action it performs, this muscle is also known as the "stretcher." (It should be kept in mind that the suspensory muscles also play an important part in stretching the larynx antagonistically between several points, and so assist the stretchers in producing the right amount of tension in the vocal folds.)

The vocal folds themselves are also important active components in the working of the larynx (Fig. 19). The vocal folds are comprised of complex fibers that can tense in various directions; they are covered by a thin membrane, the vocal bands. (These are what can get damaged if subjected to too much air pressure, as in rock and roll singing or chronically bad speaking, causing "nodes" on the vocal cords.) The vocal folds are actually quite complex, being able to tighten in some places and not in others, and being responsible for all kinds of subtle nuances in vocal production. They are made up of two muscles, ***thyroarytenoid*** and ***vocalis,*** and are also known as "tensors" of the throat—"throat" in this context referring to the larynx.

For convenience, then, the intrinsic muscles of the larynx can be divided into three categories: those which open and close (or abduct and adduct) the glottis; those which regulate the tension, or stretch, on the vocal folds (the stretchers); and the vocal folds themselves (the tensors). The muscles that rotate the pyramid cartilages are the ones that open and close the glottis. The muscles that rotate the ring and shield cartilages are responsible for regulating the tension of the vocal folds by stretching them. And the vocal folds are capable of tightening by their own action during vocalization.

It is interesting to note that none of these muscles can be consciously controlled the way we can control the finger muscles or even the muscles of the palate; they are indirectly controlled by the "ear," the part of the brain that hears sounds. When we think of a particular pitch or sound, as we do when we sing, this thought regulates the muscles of the larynx that are responsible for producing that sound. A singer is also able to activate the suspensory muscles that support the larynx, as well as the shape and condition of the pharynx, which has a very marked effect on timbre, resonance, and vocal range. Vocal training consists, to a large extent, in learning to activate this muscular system, and in being able to distinguish the various qualities required in singing. This makes it possible to regulate the vocal functions corresponding to these qualities.

SPINE AND TRUNK REGION

11. Anterior Muscles of the Cervical Spine

In the sections dealing with the head/neck region, we looked at muscles relating to the larynx and throat, which include the tongue and palate. There are several other muscles on the front of the neck, lying rather deep and connecting to the head and spine, which are not related so much to the structures of the throat and vocalization but to posture and balance (Fig. 20). Let's take a quick look at these before looking at the spine and its supporting musculature.

Longus capitis originates at the four vertebrae just below the atlas and axis (the third, fourth, fifth, and sixth) and attaches to the occiput just in front of the spine. "Capitis" means "of the head."

Longus colli, which means "long muscle of the neck," is in three sections, connecting the atlas with the upper cervical vertebrae, and the upper thoracic with the lower cervical vertebrae. In older anatomy texts, "colli" is often used to designate muscles of the neck region; "cervicis," corresponding to "cervical," is now used more commonly.

Rectus capitis anterior originates at the transverse process of the atlas and attaches onto the occiput just behind longus capitus. *Rectus capitis lateralis* originates at the transverse process of the atlas and inserts onto the occiput just to the side of the occipital condyle. (These muscles are called *sub-occipital* muscles because they are under the occiput. There are four other sub-occipital muscles on the *back* of the spine, which we'll look at when we discuss the muscles of the back.) All of these deep postural muscles on the front of the spine are important in maintaining the length and support of the upper spine, and they work in conjunction with the postural muscles on the back of the spine, which we'll talk about in the sections dealing with the musculature of the back.

Fig. 20. Anterior muscles of cervical spine

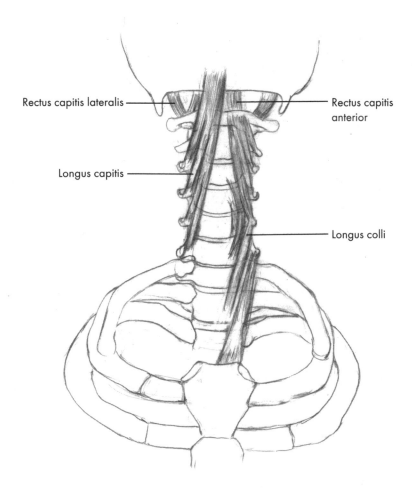

Rectus capitis lateralis

Rectus capitis anterior

Longus capitis

Longus colli

Another important group of muscles attaching to the front of the spine is the ***scalene*** muscles, which attach the middle and lower vertebrae of the neck to the upper two ribs, acting as tensile members or guy wires supporting the upper rib cage, and also aiding in inspiration. We will discuss the scalene muscles in more detail in Chapter 19, "Suspensory Muscles of the Thorax."

71

12. The Vertebrae of the Spine

The spine, or backbone, is a flexible column formed by a series of bones called vertebrae (from *vertere,* to turn) (Fig. 21). In all vertebrates, the spine forms the central support structure for the body, and it is supported or acted upon by a large and complex network of muscles. In improving one's patterns of movement and "use"—particularly when one is concerned with releasing unnecessary and harmful muscular tension—it is easy to emphasize the role of muscles and to underestimate the importance of bones, and the spine in particular. But we have to remember that in order for movement to take place, and in order to be supported against gravity, there needs to be a solid structure for muscles to act upon. There are of course other kinds of muscle in the body like the heart or the muscles lining the digestive tract, which are not related to skeletal movement and do not require bones to function. But muscles that relate to movement, which are the kind that we're interested in, act on bones and so do not exist without the skeleton. And the spine is the central core of the skeletal structure.

The spine is made up of thirty-three vertebrae in all—seven in the cervical (neck) region, twelve in the thoracic (chest) region (also called the "dorsal" region), five in the lumber (lower back) region, five in the sacral (pelvic) region, and four in the coccygeal (tailbone) region. The sacral and coccygeal vertebrae are almost undetectable, being fused to form the sacrum and coccyx, so that for all intents and purposes the spine consists of twenty-four movable vertebrae, plus the sacrum and tailbone. The tailbone, or coccyx, which is Greek for "cuckoo bird," was so named because of its resemblance to the beak of this bird; it is vestigial, being the remnant of the tail.

The main purpose of the spine is to bear weight and to provide a support structure for muscles to act upon in producing upright support and movement.

Fig. 21. Vertebral column

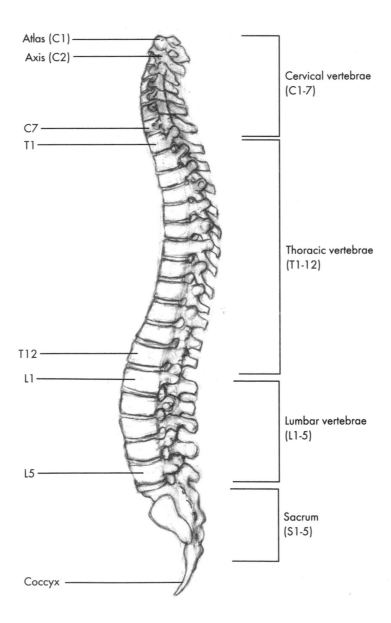

Atlas (C1)

Axis (C2)

Cervical vertebrae
(C1-7)

C7

T1

Thoracic vertebrae
(T1-12)

T12

L1

Lumbar vertebrae
(L1-5)

L5

Sacrum
(S1-5)

Coccyx

If you look at a skeleton, you'll see that the vertebrae that make up the spine have two sections (Fig. 22). At the front, the vertebrae are round, with discs in between, forming a column of vertebrae stacked on top of each other. In the back, the vertebrae form a continuous bony structure that runs up the length of the back, with various projections and protrusions.

The front, round part of the vertebra forms the weight-bearing part of the spine and is called the ***body,*** or ***centrum,*** of the vertebra. The discs in between the vertebral bodies are resilient and designed to absorb shock, and together with the vertebrae form a solid, flexible column for supporting the head and trunk. The vertebral bodies and the intervertebral discs are smaller at the top and get much larger and stronger at the bottom, where they have more weight to support.

The back part of the vertebra, called the ***arch,*** has several functions. By coming into contact with each other at the back, the vertebrae form joints which make it possible to move the spine. (When we place two vertebrae together and they fit like pieces of a puzzle, it is these joints that are fitting together.) The projections, or "processes," in this part of the vertebra, which protrude down the length of the backbone and are so prominent in dinosaurs and large mammals, form attachments for muscles and ligaments which act upon and support the spine. Finally, the back part of the vertebra forms a hole, or canal, just behind the body of the vertebra which protects the spinal cord running down the length of the spine. Branches of the spinal cord project laterally out small holes between the vertebrae right down the length of the spine, sending nerves to various parts of the body corresponding to the different levels of the spine.

The arch of the vertebra is made up of several sections. Extending from the back part of the body of the vertebra are the ***pedicles,*** which form the sides of the spinal canal. Continuing around from the pedicles is the ***lamina,*** which is a flat sheet of bone that joins the pedicles together to form the space for the spinal canal. This space is called the ***vertebral foramen***.

Fig. 22. The vertebrae and spine

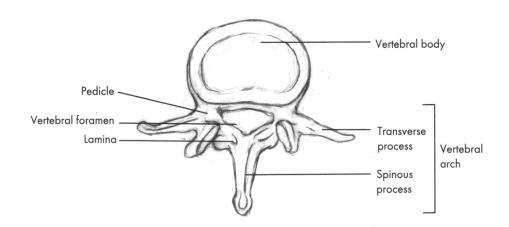

Vertebral body

Pedicle

Vertebral foramen

Lamina

Transverse process

Spinous process

Vertebral arch

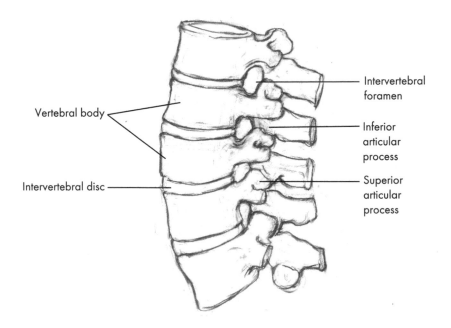

Vertebral body

Intervertebral disc

Intervertebral foramen

Inferior articular process

Superior articular process

Projecting sideways from the pedicle is the ***transverse process***—the lateral projections that form attachments for ligaments and muscles. Where the lamina come together in back, the bone projects backward and downward to form the ***spinous process***—the prominent projection at the back of the spine that can be felt underneath the skin as a series of bumps down the length of the spine. When you feel someone's spine, it is these projections you feel, not the main part of the vertebrae. These projections also provide attachments for ligaments and muscles.

Extending upward and downward from the pedicles and lamina are projections that form joints with the vertebrae above and below. Extending upward from the pedicles is the ***superior articular process***—the joint that articulates with the vertebra above it. Projecting downward from the lamina is the ***inferior articular process***—the projection that articulates with the vertebra below it. The upward projection of one vertebra fits in neatly with the downward projection of the one above it, in this way forming a series of joints along the entire length of the spine.

Between the pedicles of the vertebrae are gaps, called ***intervertebral foramina;*** branches of the spinal nerve pass laterally through these gaps down the entire length of the spine.

To summarize, the vertebra is composed of the body and the arch. The body is the round part of the vertebra which forms the weight-bearing pillar of the spine. Between the body of each vertebra is an intervertebral disc, which is a resilient and pliable structure designed to absorb shock, as well as to twist and bend. (We'll look more at the intervertebral discs when we look at the spine as a whole.) Extending backwards from the body of the vertebra are the pedicles and lamina which protect the spinal cord; the transverse processes and the spinous process that form attachments for muscles and ligaments; the superior and inferior articular processes that form joints, or points of communication, with the vertebrae above and below; and the intervertebral foramina, the small openings between the pedicles on each side of the

Fig. 23. Atlas and axis (C1 and C2)

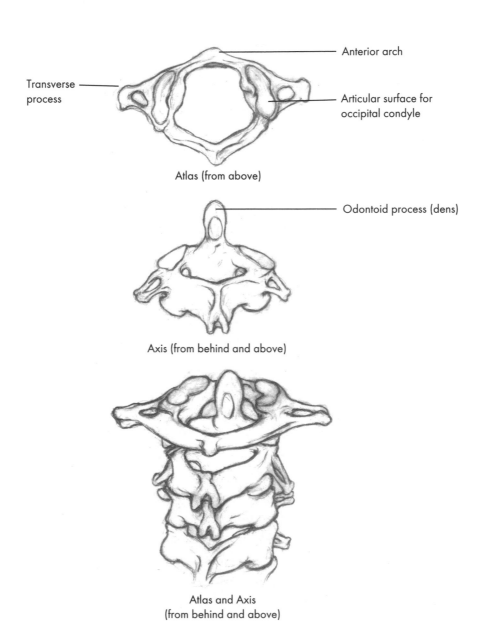

Transverse process

Anterior arch

Articular surface for occipital condyle

Atlas (from above)

Odontoid process (dens)

Axis (from behind and above)

Atlas and Axis
(from behind and above)

vertebrae through which the nerve branches pass. All of these structures comprise the arch of the vertebra.

It is important to keep in mind that not all the vertebrae are the same. We've already seen that the vertebrae and the discs at the bottom of the spine are larger than the ones above. There are other more subtle differences in vertebrae in different parts of the spine; observing an actual model gives a general idea of some of these differences.

In particular, there are two vertebrae which are highly unusual and which deserve special attention—the ***atlas*** (which is the first cervical vertebra on which the skull rests) and the ***axis*** (which is the second cervical vertebra) (Fig. 23).

If you look at the ***atlas*** (which is so named because it supports the globe of the head), there are two noticeable things about it. First, unlike all the other vertebrae, it has no body, but looks more like a ring with two concave depressions on either side of the hole in the ring. That's because this vertebra, being the topmost one, doesn't need a big wide area to support a vertebra on top of it, but instead must support the skull. As we've seen, the skull has two rounded bumps on either side of the foramen magnum, called the occipital condyles. These two bumps fit neatly into the two concave areas on either side of the ring of the atlas, which are, to quote *Gray's Anatomy,* "admirably adapted to the nodding movements of the head." This joint, which as we've seen is called the ***atlanto-occipital joint*** because it is formed by the junction of the occiput of the skull with the atlas, is the crucial point at which the head is poised on top of the spine (Fig. 24). As we saw earlier, the head tends to nod forward at this point, since its center of gravity is forward of this point of balance. This tends to counteract the pull of muscles at the back of the neck and has the effect of lengthening the spine.

The second unusual feature of this vertebra is that the transverse processes extend unusually far out to the sides—so far, in fact, that they almost seem to span the width of the occiput itself. These are important

Fig. 24. The skull and head/neck joints

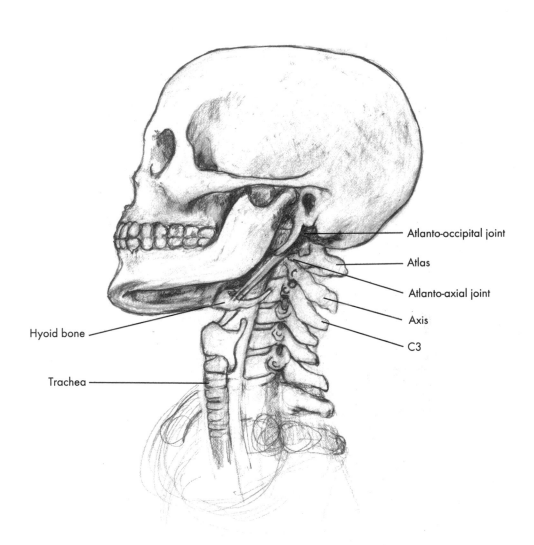

Atlanto-occipital joint

Atlas

Atlanto-axial joint

Axis

C3

Hyoid bone

Trachea

because they provide attachments for the deep sub-occipital muscles that nod and rotate the head, and that play such a crucial role in the postural support of the body against gravity. We'll look at these muscles later in more detail.

The *axis,* which is so named because it forms the pivot, or axis, upon which the first vertebra rotates with the head sitting on it, is also unusual in its structure. At the back of the axis is an upward projection called the ***odontoid process,*** or ***dens*** (meaning "tooth"). It extends up within the anterior arch of the atlas; the atlas, with the head sitting on it, rotates around the odontoid process and glides on the axis as it rotates, making it possible to rotate the head in relation to the spine. This joint where the atlas rotates on the axis is called the ***atlanto-axial joint*** because it is formed by the atlas and the axis (Fig. 24); along with the atlanto-occipital joint, it forms part of that crucial area in all vertebrates where the head articulates with the spine.

So the skull can nod on the atlas, and it can rotate in relation to the spine because the first vertebra, with the skull sitting on top of it, can pivot on the axis. In this sense, there are two head/neck joints, not one: the joint formed by the atlas and the occiput of the skull, and the joint formed by the atlas and axis. Both of these joints are reinforced by several ligaments, so that although the head is balanced up on top of the spine, it is also quite stable at this point, and limited in its movements by ligaments and muscles.

Those are some of the important bony features of the spine. We'll look next at the supporting structures of the spine that bind it into one unit—the ligaments.

13. The Spine and Its Supporting Ligaments

We saw in Chapter 12 that the vertebra has two parts: the body, which forms the supporting column of the spine, and the arch, which forms a canal down the back of the spine for protecting the spinal cord. The arches also form joints so that the vertebra are linked together, and points of attachment for ligaments and muscles that strengthen and support the spine, as well as openings for the outlet nerves which emerge from the spine laterally down its length. The spine is made up of twenty-four movable vertebrae and the sacrum and coccyx; the first two vertebrae, the atlas and the axis, form the joints that articulate the head with the spine.

A number of ligaments hold the vertebrae together, forming a trusswork between the vertebrae and strengthening the spine as a whole (Fig. 25). The *interspinous ligament* runs between the spinous processes, firmly binding them together. The *supraspinous ligament* runs down the outside of the spinous processes, linking the spinous processes together right down the length of the back. The *intertransverse ligament* connects the transverse processes of the vertebrae to each other. There is a fourth ligament, *ligamentum flava,* which runs between the lamina and which is elastic because it connects the movable joints of the spine; its function is to help extend the spine.

Two long ligaments support the bodies of the vertebrae right down their length. These ligaments are very strong and bind the spine into one flexible, weight-supporting unit. The *anterior longitudinal ligament* runs along the front of the bodies of the vertebrae. The *posterior longitudinal ligament* runs along the back of the bodies of the vertebrae right inside the spinal canal. There are also four ligaments supporting the top two vertebrae in their connection with the skull (not pictured). Finally, the *ligamentum nuchae*

Fig. 25. Ligaments of the spine

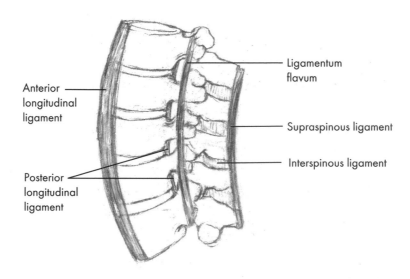

Anterior longitudinal ligament

Ligamentum flavum

Supraspinous ligament

Interspinous ligament

Posterior longitudinal ligament

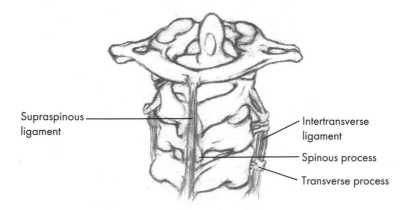

Supraspinous ligament

Intertransverse ligament

Spinous process

Transverse process

("nuchae" is another word for "neck") connects the skull with the spine. You can feel this become taut if you drop your chin to your chest and feel just below the occiput.

Between the bodies of the vertebrae, as we saw earlier, are the ***intervertebral discs*** (Fig. 22). The discs are made up of fibrous and gelatinous material which is highly absorbent and resilient. In addition to being capable of absorbing shock, the intervertebral discs make it possible to bend the spine. They are wedge-shaped, giving the spine its natural curves (the spine is not supposed to be straight, but has curves built into it).

The intervertebral discs are composed of tough, fibrous outer shells encasing a nucleus filled with fluid. The nucleus is highly absorbent but, when subjected to constant pressure, can lose fluid over time, which accounts for the variability in height that can occur from morning to night, or with in astronauts in space—sometimes up to an inch or more. With age, the discs gradually lose their elasticity and ability to imbibe moisture, which partially accounts for loss of height in old age.

When the intervertebral discs are subjected to constant pressure by exaggerated bending and postural distortion, this compresses and pinches the back portion of the disc, causing it to rupture (Fig. 26). It then extrudes fluid, which can impinge on the spinal cord or one of its outlet nerves. This condition, which is often called "sciatica," is sometimes attributed to a "slipped disc," which is actually a misnomer; discs can't slip, being firmly attached to the vertebrae above and below. The real cause is actually a herniation, or rupture, that occurs in the wall of the disc, causing a leakage of fluid and a bulge in the disc, which then impinges on the nerve.

This condition is prone to occur at the lower back, particularly at the disc between the first sacral and fifth lumbar vertebrae. Because of the angle of the sacrum, the fifth lumbar vertebra exerts a shearing pressure on the disc between L1 and S5, pinching the back of the disc. There are various ligaments

Fig. 26. Lower spine showing pinched disc

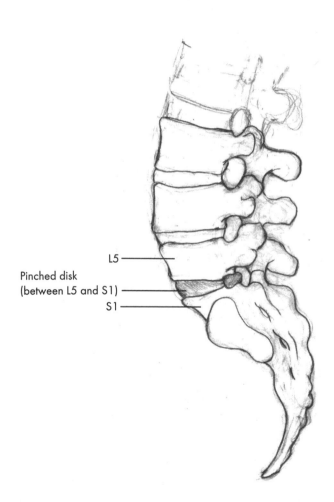

L5

Pinched disk
(between L5 and S1)

S1

which support this "lumbosacral" joint and which are designed to prevent this from happening (we'll look at these when we get to the pelvis), but constant pressure often causes problems in this region of the back.

So much for the specific components of the spine. Let's look now at the spine as a whole. Many people think of the spine as having three curves—the cervical, thoracic, and lumbar—when in fact there is a fourth—the final lower curve formed by the sacrum and coccyx (Fig. 21). There is a definite reason for this. In a four-footed animal, the sacral and thoracic curves form one large arching structure that functions like a bridge to support the body over the fore and hind limbs. Additionally, there is another concave curve at the neck, where the head extends beyond the spine. In the erect human posture, this cervical curve provides a backward curve to support the head vertically on the spine. However, the broad curve of the back cannot support erect posture, as we can see in a young infant, whose body will buckle into a foetal curve if you try to sit it upright too soon. In order to sit or stand erect, another compensating curve must develop in the lumbar region to counterbalance the thoracic curve. The result is four curves—cervical, thoracic, lumbar, and sacral (two inner ones balancing two outer ones)—which provide a flexible, balanced column that can support upright posture.

We normally associate movement with action of the limbs at the joints, but of course the spine is capable of movement as well. The main movements we can make at the spine are extension, flexion, lateral flexion, and rotation— that is, we can bend forward and back (flexion and extension); we can bend sideways (lateral flexion); and we can rotate the spine. The spine is designed to move in these ways, the intervertebral discs compressing and then rebounding from the pressure to allow movement at the vertebral joints. It is only when pressure from constant bending and postural deformation is too great that the discs lose the ability to rebound.

Because the spine is subject to so many stresses, the vertebrae can easily become displaced, or slightly dislocated, which in turn can impinge on the spinal cord and impair its ability to conduct nerve impulses. When chiropractors make adjustments, they are correcting these dislocations, or "subluxations," as they are sometimes called, with the goal of restoring the ability of the nerves to conduct impulses. However, when the spine has its natural length and the proper support of its surrounding muscles, this is of course the most effective way to ensure that the vertebral joints remain healthy and properly aligned.

14. Muscles of the Back: Deep Layers

To review the spine, we saw that there are twenty-four movable vertebrae, connected by a series of short ligaments at the back and supported by two long ligaments in front and back. The skull sits on the top vertebra, or atlas, forming the atlanto-occipital joint; the atlas, with the skull sitting on it, rotates on the second vertebra, the axis.

Let's turn now to the muscles of the back and spine. If you look in *Gray's Anatomy,* you'll see that there are no fewer than five layers of back muscles. The most superficial layer is made up of two large, powerful muscles many of us are familiar with—***latissimus dorsi*** and ***trapezius***. There are muscles in the middle layers which tend to run horizontally and whose function is to support the scapulae and ribs. And there are deeper longitudinal layers lying right along the spine. The top layers tend to be involved in larger, powerful movements; the deeper layers are mainly postural in function.

Because anatomy is done by dissection, the outer, superficial layers of muscle are usually looked at first, and then these layers are laid open to reveal the deeper layers. We'll begin, instead, with the deeper muscles first, and from there look at the more superficial muscles on top of them. There are two layers of deep muscles of the back, and these muscles, which as I said are mainly postural in function, form the extensor system of the back—the muscles that support us in the erect posture. The first layer is comprised of a series of small muscles supporting the spine along its entire length; some of these correspond to the short ligaments that bind each vertebra to the one above and below.

The ***transversospinalis*** muscles include three groups—multifidus, semispinalis, and rotatores (Fig. 27). ***Multifidus,*** which can be found along most of the length of the spine, runs obliquely from the transverse process of one

Fig. 27. Back muscles: 1st layer (transversospinalis muscles)

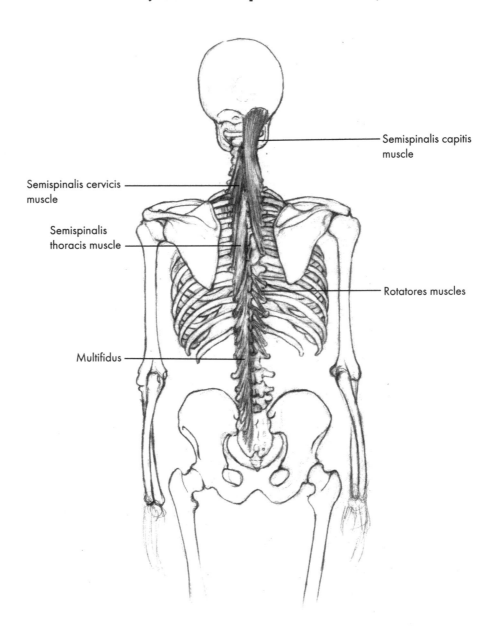

Semispinalis cervicis
muscle

Semispinalis
thoracis muscle

Multifidus

Semispinalis capitis
muscle

Rotatores muscles

vertebra to the spinous process of the vertebra above it, filling up the groove on either side of the spinous processes of the vertebrae. ***Semispinalis thoracis*** continues this pattern in the thoracic region, but overlapping more than two vertebrae at a time. ***Semispinalis cervicis*** continues this pattern in the neck region; and ***semispinalis capitis*** ("capitis" means "of the head") originates from the transverse processes of the cervical vertebrae and attaches to the occiput. Semispinalis capitis is sometimes called "complexus."

Also part of the transversospinalis group, ***rotatores*** connects the transverse process of one vertebra to the lamina of the vertebra above it; this muscle assists in rotation, which occurs mainly in the thoracic region of the spine.

Three other deep muscles of the spine comprise the deepest layer of muscles of the back (Fig. 28). In the last section we saw that the intertransverse ligament connects the transverse processes of the vertebrae. The ***intertransverse*** muscle corresponds to these ligaments, connecting adjacent transverse processes of the vertebrae. The ***interspinalis*** muscle connects adjacent spinous processes, and so corresponds to the interspinous ligament (Fig. 28).

Finally, ***levator costae*** and ***levatores costarum*** originate at the transverse processes of each vertebrae and, running obliquely downward and outward, attach to the ribs below. Levator costae is a single muscle; levatores costarum, found at the lower thoracic vertebrae, is comprised of a short and a long muscle, the shorter one attaching to the rib below, and the longer attaching two ribs below its origin. Since these muscles are elevators of the ribs, we'll look at them again when we discuss the thorax and ribs.

So the deepest layer of back muscles, which lie along the vertebrae of the spine, is comprised of multifidus, semispinalis thoracis, semispinalis cervicis, semispinalis capitis, and rotatores, which make up the transversospinalis group. Levator costae and levatores costarum attach to the transverse processes of each vertebra and, running obliquely downward, attach to the ribs below. And the intertransverse and interspinalis muscles also connect the processes of the vertebrae.

Fig. 28. Back muscles: 1st layer (cont.)

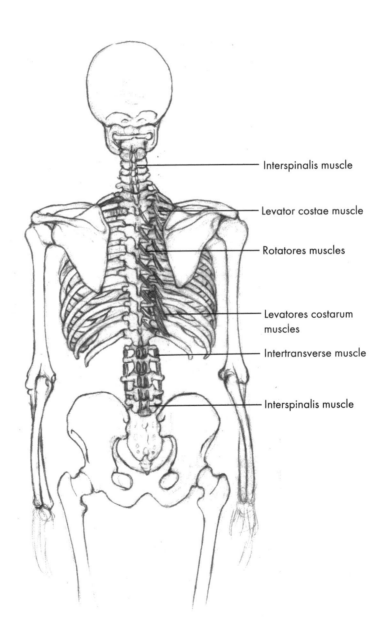

Interspinalis muscle

Levator costae muscle

Rotatores muscles

Levatores costarum muscles

Intertransverse muscle

Interspinalis muscle

These deep muscles of the spine form an essential part of the network of postural muscles that maintain erect posture. They not only aid in bending and rotation of the spine—in other words, in active movement—but also in maintaining the support of the vertebral column. Beginning with the sacrum as a kind of fixed point of support, the fibers of these deep muscles, acting upon the lumbar vertebrae, help to support the spine in this region; the lumbar vertebrae then become a fixed point for the muscles to move the thoracic vertebrae, and so on up the length of the spine, so that the small muscles of the back, with the help and support of the ligaments, have the overall effect of straightening the spine and maintaining its length or "elongation." It is also interesting to note that, along with the large extensor muscles of the back, these deep postural muscles are designed to alternately contract and relax throughout the length of the back, so that some fibers have a chance to rest while other fibers contract, making it possible to maintain postural support without any muscular fatigue—provided we don't interfere with the length of the spine so essential to the proper functioning of this system of muscles!

There is one more very important group of muscles that we must look at in order to complete the deep layer of extensor muscles of the spine, two of which I mentioned earlier. These are the **sub-occipital** muscles, so called because they lie under the occiput (Fig. 29). When we discussed the muscles attaching to the base of the skull, we saw that the extensor muscles that run down the back are all related to the base of the skull and head balance. The sub-occipital muscles are continuous with this deep layer of small muscles of the back; they relate the skull to the two topmost vertebrae of the spine.

There are actually six sub-occipital muscles. We saw earlier that **rectus capitis anterior** originates at the transverse process of the atlas and attaches onto the occiput just behind longus capitis. We also saw that **rectus capitis lateralis** originates at the transverse process of the atlas and inserts into the occiput next to the occipital condyles (see Fig. 20). Both of these muscles

Fig. 29. The sub-occipital muscles

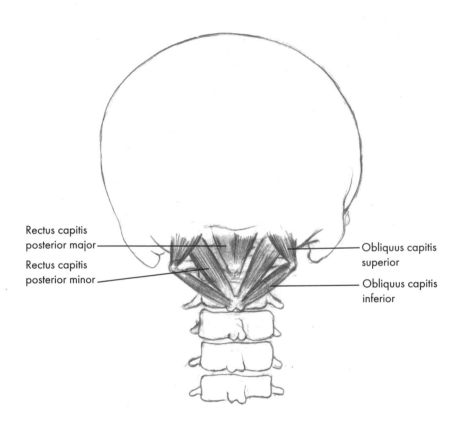

Rectus capitis posterior major

Rectus capitis posterior minor

Obliquus capitis superior

Obliquus capitis inferior

attach to the base of the skull anterior to the spine and so are not part of the extensor muscles in back.

Four more sub-occipital muscles attach to the base of the skull just *behind* the spine (Fig. 29). ***Rectus capitis posterior major*** connects the spinous process of the axis, or second cervical vertebra, to the occiput. ***Rectus capitis posterior minor*** originates at the back of the atlas and attaches onto the occiput. ***Obliquus capitis inferior*** originates at the spinous process of the axis and passes obliquely outward to insert into the transverse process of the atlas. ***Obliquus capitis superior*** arises from the transverse process of the atlas and passes obliquely upward to insert into the occiput.

Again, these small muscles, which complete the sub-occipital group, relate the first two vertebrae to the balance of the head. When the head is pulled back and down by a shortening of the larger skeletal muscles of the neck, the shortening in the larger muscles interferes with the forward balance of the head that exerts stretch on these small muscles. When we bring about release and lengthening in the muscles of the neck and back, this allows the head to go forward and up, and these deeper muscles begin to sensitively register changes in the balance of the head, which activates the postural muscles all along the spine. The sub-occipitals therefore form an essential part of the postural system, relating the balance of the head to the extensor muscles running the length of the spine.

The second layer of back muscles is called ***sacrospinalis,*** or ***erector spinae,*** and is responsible for maintaining erect posture (Fig. 30). This is an extensive system of muscles which, beginning at the sacrum, form overlapping bundles of muscles running the entire length of the spine in three columns. It is these muscles that elastically stretch when we lie in semi-supine, or the "constructive rest position," giving us the feeling that the back is a broad and continuous system of elastic muscle that connects from the sacrum right up to the head. (The two muscles that we can feel bulging in the

Fig. 30. Back muscles: 2nd layer
(sacrospinalis or erector spinae)

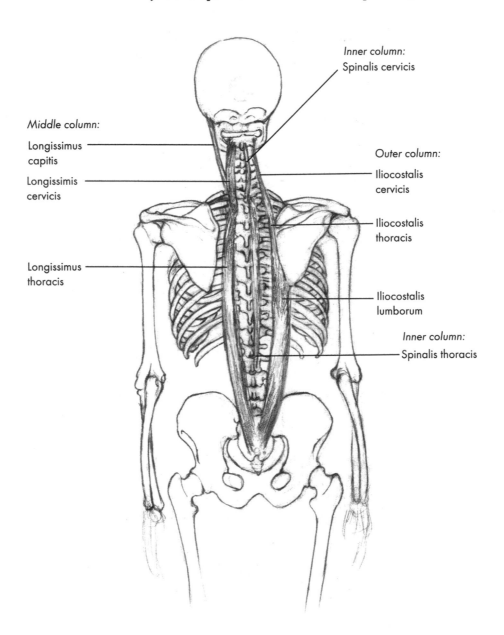

Inner column:
Spinalis cervicis

Middle column:
Longissimus
capitis

Longissimis
cervicis

Longissimus
thoracis

Outer column:
Iliocostalis
cervicis

Iliocostalis
thoracis

Iliocostalis
lumborum

Inner column:
Spinalis thoracis

lower back are part of this sacrospinalis mass; it is these muscles that are most prone to go into spasm when the back muscles are shortened and overworked.)

To simplify this group of muscles, think of it in three columns. The outer column, *iliocostalis,* originates at a broad attachment at the sacrum which fans up and out to attach to the lower ribs. This muscle is then continued by two more sections that originate at the lower ribs and attach to the upper ribs, and from the upper ribs to the transverse processes of the cervical spine.

The middle column, *longissimus,* runs more directly upward to the transverse processes of the lumbar and thoracic vertebrae, forming the longest section of the sacrospinalis mass. This muscle is continued upward in two more bundles originating at the transverse processes of the thoracic vertebrae and attaching to the cervical vertebrae, and originating at the cervical vertebrae and attaching to the mastoid process of the skull.

The inner column, *spinalis,* originates at the spinous processes of the lumbar and thoracic vertebrae and attaches to the spinous processes of the upper thoracic vertebrae. This muscle is continued upward by another section that originates at the lower cervical vertebrae and attaches to the upper cervical vertebrae.

So there are three bundles of muscles—iliocostalis, longissimus, and spinalis—that comprise the sacrospinalis group. Each one has different sections, each section being named for the region it occupies. Iliocostalis has three sections: iliocostalis lumborum, iliocostalis thoracis, and iliocostalis cervicis. Longissimus has three sections: longissimus thoracis, longissimus cervicis, and longissimus capitis. And spinalis has two sections: spinalis thoracis and spinalis cervicis.

Originating at the sacrum as the stable starting point, these three columns of muscle run upward to support the lower region of the back and spine, then begin slightly below this to support the spine at the upper back, and finally, begin just below this point to support the neck and head, forming an extensive sheet of muscles that, beginning at the sacrum and leapfrogging upward, maintain erect posture. Although gravity acts in a downward direction, the weight of the body which lies in front of the spine (as well as any weight we might carry in our arms) constantly buckles us forward; these extensor muscles counteract this tendency and so maintain the torso and spine in the erect posture. Together with the small muscles linking the vertebrae, which maintain the elongated support of the spine itself, these muscles form the extensor system of the back, which, acting on the spine and ribs and connecting right up to the head, work as one functional system to maintain erect posture. When you lie in the semi-supine position, one of the main things you are aiming at is elongation of the spine, giving the sacrospinalis sheet a chance to lengthen and release, which is the condition under which these muscles of the back and spine function most efficiently.

15. Muscles of the Back: Middle and Superficial Layers

We've now looked at the extensor muscles that comprise the first two layers of the back musculature—the deeper muscles supporting the vertebrae of the spine and maintaining the erect posture. The first layer is comprised of all the small muscles connecting each vertebrae along the entire length of the spine up to the occiput, including the sub-occipital muscles connecting the occiput with the first two vertebrae. On top of this is the second layer—the large and extensive sacrospinalis group which, beginning at the sacrum, runs up the back in several columns, ultimately attaching to the base of the skull. These two layers—the deep layer along the vertebrae that includes the sub-occipitals, and the sacrospinalis group—form the extensors of the back, the postural muscles which help to maintain erect posture and which are designed not to fatigue.

Let's continue now with the middle, or third and fourth, layers of back muscles which, unlike the first and second layers, are involved not so much in erect posture, but in supporting the ribs and scapulae.

The third layer is comprised of the following four muscles: ***serratus posterior superior, serratus posterior inferior, splenius capitis,*** and ***splenius cervicis*** (Fig. 31).

Serratus posterior superior originates at the spinous processes of the last cervical and first two thoracic vertebrae and, inclining downward and outward, inserts like four fingers into the second, third, fourth, and fifth ribs.

Serratus posterior inferior originates at the spinous processes of the upper lumbar and lower thoracic vertebrae, passes obliquely up and outward and, breaking like serratus superior into four branches, inserts into the four lower ribs. The word "serratus" means "serrated" and is descriptive of this muscle's appearance. It is called serratus *posterior* because there is a serratus *anterior* muscle on the front of the body.

Fig. 31. Back muscles: 3rd layer

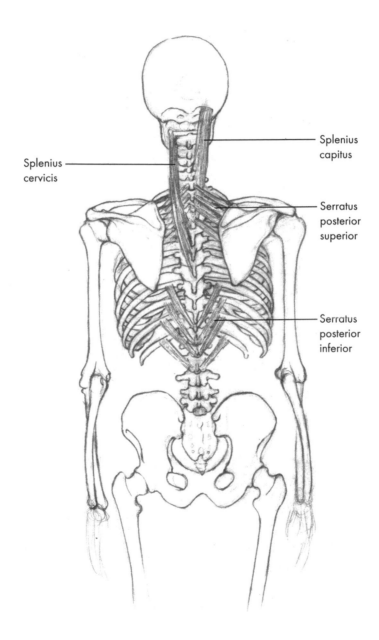

Splenius cervicis

Splenius capitus

Serratus posterior superior

Serratus posterior inferior

The serratus muscles act upon the ribs, serratus superior raising the ribs, and serratus inferior drawing the lower ribs down and widening the back. Along with quadratus lumborum (which attaches from the pelvis to the lowest rib and which we'll discuss when we get to the thorax) and levatores costarum (which we've seen is one of the deeper muscles attaching to the transverse processes of the ribs), serratus posterior plays an important role in freeing the ribs and widening the back. When they are releasing, these muscles contribute to the freedom and fullness of the lower back, as well as the proper working of the diaphragm, which is directly related to the lower ribs.

So although the lengthening of the back tends to bring about widening, the muscles that widen the back and the muscles that lengthen the back more or less comprise different muscle layers and correspond to different functions. The lengthening of the back corresponds to the deep extensor muscles of the back, which form a vertical system of muscles. The width of the back corresponds more to the muscles attaching to the ribs and scapulae, which tend to run more obliquely and horizontally, and which relate primarily to the support of the lower and upper ribs, the action of the diaphragm, the elastic support of the lower back, and the support of the shoulder girdle—again, very different functions than those performed by the deep extensor muscles of the back.

The *splenius* muscle (meaning "a bandage") originates at the upper thoracic and lower cervical vertebrae and the ligamentum nuchae and, fanning obliquely upward, attaches to the transverse processes of the upper cervical vertebrae *(splenius cervicis)* and the mastoid process *(splenius capitis)*. Along with other muscles that insert into the occiput, splenius pulls the head back, is involved in rotation of the head, and supports the head on the spine.

The fourth layer is comprised of ***levator scapulae, rhomboid minor,*** and ***rhomboid major*** (Fig. 32). Originating at the transverse processes of the atlas and upper cervical vertebrae, ***levator scapulae*** attaches to the side of the scapula. This muscle, as its name suggests, elevates the scapula.

Fig. 32. Back muscles: 4th layer

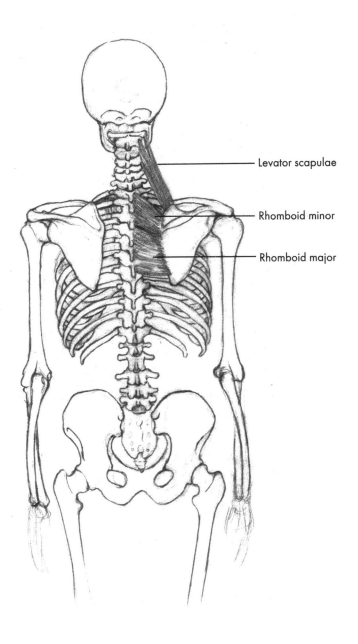

Levator scapulae

Rhomboid minor

Rhomboid major

Rhomboid minor originates at the seventh cervical and first thoracic vertebrae and passes downward and outward to insert into the spine of the scapula. *Rhomboid major* originates at the spinous processes of the four or five upper thoracic vertebrae and attaches to the lower spine of the scapula. Both of these muscles, which are so named because of their rhomboidal shape, act upon the scapula and help to stabilize it when the arms are being moved.

The fifth, or superficial, muscle layer of the back consists of two powerful sheets of muscle covering most of the back, *trapezius* and *latissimus dorsi* (Fig. 33). *Trapezius* (trapezium, a four-sided figure) is a diamond-shaped muscle covering the neck and upper shoulders. It originates at the occiput, the ligamentum nuchae, and the spinous processes of the seventh cervical and all the thoracic vertebrae. From this wide origin its fibers extend laterally to converge into the spine of the scapula, the acromion process, and the clavicle. Like serratus anterior (which we'll look at later when we discuss the shoulder girdle) and the rhomboids, trapezius acts upon the shoulder blades, retracting them or pulling them together, bracing the shoulder in strong skeletal movements, and elevating the shoulder as in lifting weights.

Latissimus dorsi, which means "broadest muscle of the back," is also a broad, flat sheet of muscle; it covers the lower half of the back and converges to insert into the humerus, the upper arm bone. It originates at the spinous processes of the sacral, lumbar, and lower thoracic vertebrae and the iliac crest, or hip bone, and from this very broad origin passes obliquely upward in a twist, like a towel that is being wrung out, to converge into the bicipital ridge of the upper humerus, just above the point where the pectoralis major also inserts. This muscle lends the torso the athletic, triangular shape seen in body builders. It is one of the lifting muscles of the back; it depresses the upper arm and stabilizes the body in relation to the arms, as when we support weight on crutches. It is also involved in powerful downward movements of the arm such as swinging an axe and is a key muscle in contributing to the widening and support of the back in relation to the upper arms.

Fig. 33. Back muscles: 5th (superficial) layer

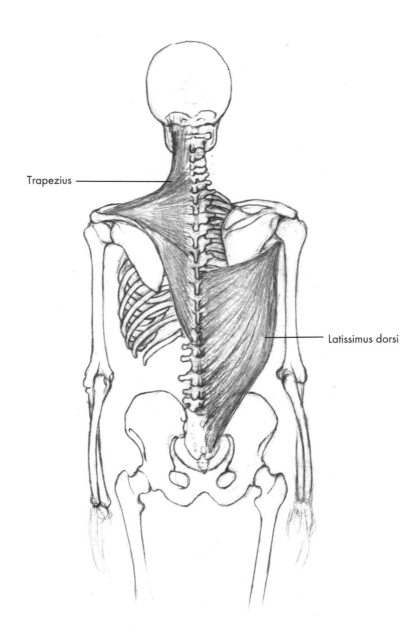

Trapezius

Latissimus dorsi

To summarize the muscle layers of the back, which reverses the order you'll find in *Gray's Anatomy:*

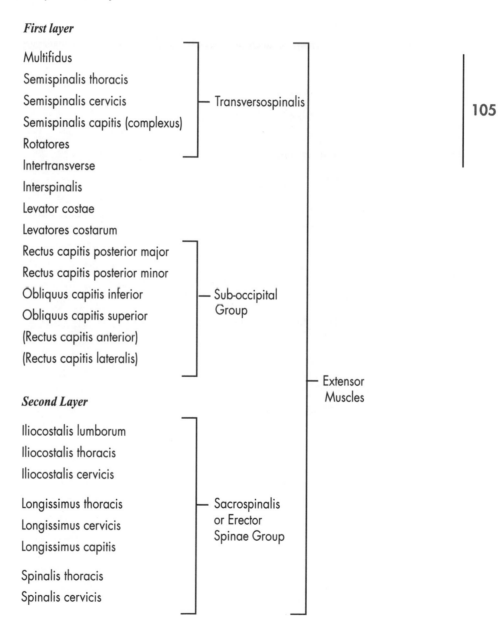

First layer

Multifidus
Semispinalis thoracis
Semispinalis cervicis
Semispinalis capitis (complexus)
Rotatores

— Transversospinalis

Intertransverse
Interspinalis
Levator costae
Levatores costarum

Rectus capitis posterior major
Rectus capitis posterior minor
Obliquus capitis inferior
Obliquus capitis superior
(Rectus capitis anterior)
(Rectus capitis lateralis)

— Sub-occipital Group

— Extensor Muscles

Second Layer

Iliocostalis lumborum
Iliocostalis thoracis
Iliocostalis cervicis

Longissimus thoracis
Longissimus cervicis
Longissimus capitis

Spinalis thoracis
Spinalis cervicis

— Sacrospinalis or Erector Spinae Group

Muscles of the back continued:

Third Layer

Serratus posterior superior
Serratus posterior inferior
Splenius capitis
Splenius cervicis

Fourth Layer

Levator scapulae
Rhomboid minor
Rhomboid major

Fifth Layer

Trapezius
Latissimus dorsi

16. Muscles Attaching to the Front of the Spine

Before going on to look at the muscles of the chest and abdomen, let's glance at the muscles lying on the front of the spine (Fig. 34). We looked earlier at the anterior muscles of the cervical spine—the deeper postural muscles linking the skull and neck, and supporting the cervical or upper spine. These muscles have a corresponding group in the pelvic region and lumbar spine. Notice that there are no muscles lying along the front of the thoracic spine, and that where there are muscles is where we tend to exaggerate the forward curves of the spine. This is because the thoracic spine requires support only in back, whereas the cervical and lumbar regions require support on both sides of the spine in order to support the head and pelvis, and in order to maintain the elongation or extension of the spine in the cervical and lumbar regions, where it tends to buckle.

To look again at the four deeper muscles lying on the front of the cervical spine, *longus capitis* originates at the four vertebrae just below the atlas and axis (the third, fourth, fifth, and sixth) and attaches to the occiput.

Longus colli, which is in three sections, connects the atlas with the upper cervical vertebrae, and the upper thoracic with the lower cervical vertebrae.

Rectus capitis anterior originates at the transverse process of the atlas and attaches to the occiput just behind longus capitis.

Rectus capitis lateralis originates at the transverse process of the atlas and inserts more laterally into the occiput.

One other group of muscles attaches to the front of the spine in the region of the neck and runs downward to attach to the upper two ribs: the three *scalene* muscles. *Scalene anterior* and *scalene medius* originate at the transverse processes of the middle and lower vertebrae of the neck and insert into the first rib. *Scalene posterior* arises from the lower cervical

Fig. 34. Muscles attaching to the front of the spine

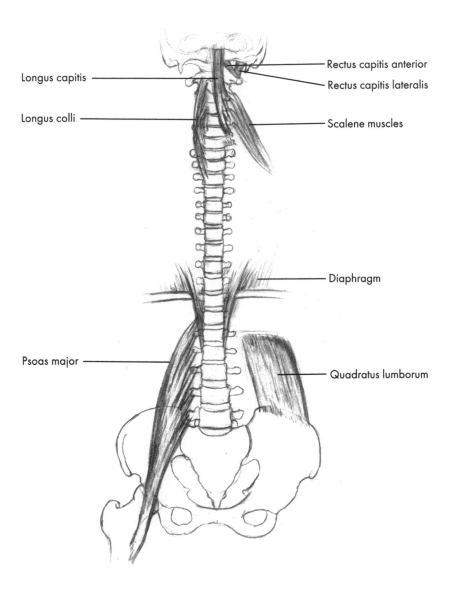

Longus capitis

Longus colli

Rectus capitis anterior

Rectus capitis lateralis

Scalene muscles

Diaphragm

Psoas major

Quadratus lumborum

vertebraeand inserts into the second rib. All three muscles support the ribs and are involved in breathing.

Corresponding to the muscles attaching to the front of the cervical spine are the muscles attaching to the front of the lumbar spine. Like the deeper muscles in the region of the neck, these muscles are also mainly postural in function, supporting the spine and pelvis and helping to maintain erect posture.

Psoas major arises from the bodies and the transverse processes of the lumbar and last thoracic vertebrae and, narrowing as it passes down across the front of the pelvis, inserts into the upper thigh bone, or femur.

Psoas minor, which is narrower than psoas major and runs on top of it, arises from the sides of the bodies of the lowest thoracic and upper lumbar vertebrae and converges into a tendon which inserts like a strap into the fascia of the ilium, or hip bone, of the pelvis.

Quadratus lumborum originates at the hip bone, or pelvic crest, and runs obliquely upward to attach to the transverse processes of the lumbar vertebrae and the twelfth, or lowest, rib.

In addition to these three muscles, the *diaphragm* attaches to the lumbar vertebrae and, fanning up and forward, forms a powerful, dome-shaped muscle whose downward contraction assists in breathing (we will look more at the diaphragm when we discuss the thorax and breathing). Like the psoas and quadratus muscles, the diaphragm exerts a strong pull on the lumbar spine, and so must be considered when we look at the muscles attaching to the lower spine.

Again, notice that there are no muscles lying on the front of the spine in the thoracic region. The thoracic spine forms the basic convex curve that supports the rib cage; no muscles attach to the front of the spine within this basic life-support structure that houses the lungs and heart. The inward curves in the cervical and lumbar regions of the spine, however, require postural support on both the back and front of the spine—precisely those points where we tend to exaggerate the concave curves of the neck and lower back, and where the pelvis must be supported in relation to the lumbar spine and the head in relation to the cervical spine. Both of these areas must be released in order to restore length to the spine and to allow the extensors of the back to support fully upright posture—the pelvic/lumbar region to allow the lower back to release and lengthen, and the throat and cervical region to allow lengthening in the neck and upper spine.

THORAX AND ABDOMEN

17. The Thorax and Muscles of Respiration

The thorax (which refers to the breastplate or armor worn by the ancient Greeks) is formed by the ribs and the sternum and has several functions (Fig. 35). First, it provides a bony protective structure for the organs of breathing and circulation—the lungs and heart. Second, its movements, along with those of the diaphragm, are responsible for breathing. Third, it provides part of the structure of attachments and supports for the back and torso muscles.

There are twelve ribs on each side of the body, which correspond to the twelve thoracic vertebrae of the spine. The first seven attach in front to the sternum, or breastbone; these are called *true* ribs. The remaining five are called *false* ribs, because they do not attach directly to the sternum, but join each other to form an arch below the sternum called the *costal arch,* which can be easily felt below the sternum. The final two ribs are called the *floating ribs,* because they do not attach in front. The ribs are not bony all the way around. Before reaching the sternum, they become cartilage, so that the connection of the ribs with the sternum is cartilaginous and quite flexible. The costal arch is also made up of cartilage.

Within the rib cage, of course, are the lungs and heart. If you identify the sternum, the heart sits right behind the lower part of that and a little to the left; the lungs are on either side of the heart. The *diaphragm* forms the lower boundary of the thorax (see Fig. 40); the heart and lungs lie above the diaphragm, and all the other major internal organs lie below the diaphragm, which forms a boundary between these upper and lower regions of the trunk. Diaphragm is actually a descriptive term which the Greeks gave to this muscle: it means "a partition wall."

We normally do not think of the ribs as having joints, but they in fact form joints with the vertebrae of the spine—called *costovertebral* joints (Fig. 36)—

Fig. 35. The rib cage

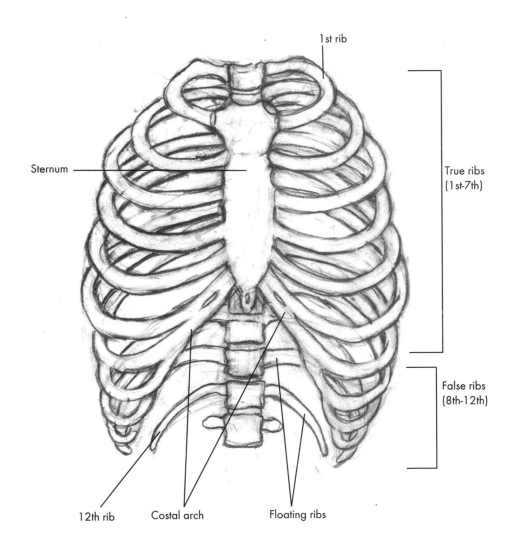

1st rib

Sternum

True ribs
(1st-7th)

False ribs
(8th-12th)

12th rib   Costal arch   Floating ribs

and are able to rotate at this point. Each rib articulates with the spine in several places. First, the head of the rib articulates with the lower part of the body of one vertebra and the upper part of the one below it, as well as with the disc in between the two vertebrae. Second, the neck of the rib articulates with the transverse process of the lower of these two vertebrae. The rib is firmly bound at each of these articulations by several ligaments, permitting a limited rotation at the joint that nevertheless translates into quite a lot of movement over the entire length of the rib. Some of the ribs have simpler articulations, but the main thing to keep in mind is that the ribs actually articulate at the spine to permit the movements essential to breathing. The cartilage in the front of the ribs is somewhat flexible and forms gliding joints in its connection with the sternum. These joints make it possible for the ribs to rise and descend during breathing, and allow movement of the ribs where they join the sternum.

The ribs are all very different in character (Fig. 35). The top rib is small, squat, and round. We often think of the upper ribs as being almost as large as the middle ones. However, this upper rib, which forms an opening into the thorax, or thoracic inlet, is quite small—only a third the diameter of the shoulder girdle. It is through this opening that the windpipe, esophagus, and other structures pass from the neck down into the chest. The next rib is larger but shaped like the first. Going downward, the ribs increase in length and begin to slant down obliquely, corresponding to the muscles of the trunk, which also tend to slant and spiral around the trunk. The final two floating ribs, which are much shorter than the ones above, are very flexible in their movements because they do not attach to anything in front; their function is mainly to provide attachments for the diaphragm.

In back, the ribs do not extend directly to the sides to form the rib cage; they actually come slightly back almost to the level of the spinous processes of the vertebrae. This means that there is a gap between the spinous processes and the posterior part of the rib on either side. This gap is filled up with the longitudinal extensor muscles, which give the back a flat appearance.

Fig. 36. The costovertebral joints

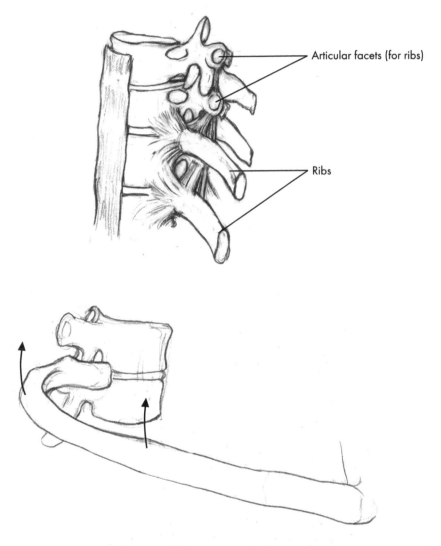

Articular facets (for ribs)

Ribs

Movement of rib at costovertebral joint

Fig. 37. Ribs during exhalation and inhalation

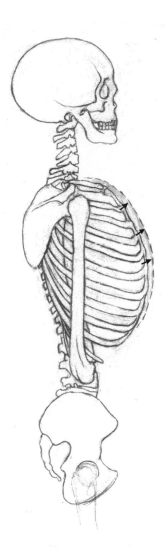

Movement of ribs

Fig. 38. The intercostal muscles

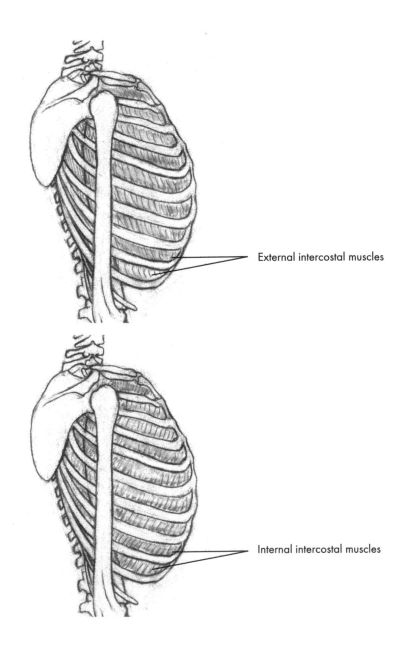

External intercostal muscles

Internal intercostal muscles

The movement of the ribs is of course crucial to breathing. When we breathe, the sternum moves slightly up and down, and the ribs, by rotating where they articulate with the spine, move like pail handles being raised slightly (Fig. 36). The raising of the ribs increases the lateral dimension of the thorax, and the raising of the sternum in front increases the antero-posterior dimensions of the thorax as well (Fig. 37). This increases the entire circumference of space within the thorax, causing air to flow into the lungs. Of course, not all the ribs move in the same way: the first ribs move very little, and there is in general more movement as you go lower down. The final two floating ribs are quite mobile, or at least should be.

When the muscles of the trunk are functioning properly, the ribs are able to move flexibly, but because of tension and postural distortion, they are typically fixed and quite rigid. When this happens, the entire thorax moves too much as a whole, and the diaphragm, the other main agent of breathing, becomes overworked to make up for the lack of movement in the ribs. Also, the rib cage as a whole becomes fixed in a wrong attitude, usually somewhat thrown backward and shortened in front and then held up, narrowing the back and preventing the free action and mobility of the ribs in breathing, as well as interfering with the general upright support of the trunk. When we are properly supported by the postural muscles and are not shortened in stature, this tends to re-orient the entire rib cage, which in turn allows a widening of the back and an increased mobility and action of the ribs.

There are two layers of rib, or intercostal, muscles which are responsible for breathing (Fig. 38). The *external intercostals* arise from the lower border of each rib and attach to the upper border of the rib below, running obliquely down and forward. Underneath this layer are the *internal intercostals,* which arise from the inner surface of each rib and slant down and back, in the opposite direction to the external intercostals, to attach to the rib below. The external intercostals function mainly to elevate the ribs, which

119

Fig. 39. Transversus thoracis

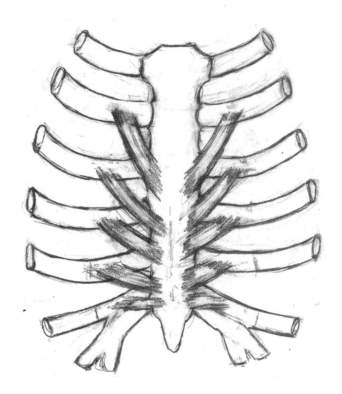

(Internal view of thoracic wall)

increases the width of the thoracic cavity and therefore causes inspiration. The internal intercostals depress the ribs when actively breathing out.

Transversus thoracis lies on the inner surface of the lower part of the sternum (Fig. 39). Its fibers extend up and out, like the splayed fingers of a hand, and insert into the costal cartilages of the second, third, fourth, fifth, and sixth ribs. This muscle, which aids in forceful expiration, is the muscle that you can sometimes feel gripping in the inner chest; it contributes to the rigidity of the chest in many people who raise and fix the chest in speech and breathing.

The *diaphragm* is the main muscle of breathing (Fig. 40). It is basically dome-shaped, formed by muscular fibers that attach all around the circumference of the lower thorax and converge upward into a central peak which is tendinous. The fibers forming this muscle arise from the lumbar vertebrae, the bottom end of the sternum, and the cartilages of the lower ribs. The diaphragm is actually a double dome, the right dome being larger than the left, so that it is lopsided; the aorta, the esophagus, and the vena cava (the large blood vessel that returns blood to the heart), as well as other structures, pass right through it to get from the chest into the abdominal region. When the fibers of the diaphragm contract—particularly the portion of the muscle attaching to the lumbar vertebrae, which pulls the central tendon of the diaphragm down—the dome is depressed, which enlarges the space inside the rib cage.

Notice where the diaphragm is in the body. If you identify in yourself where the lowest ribs are in front, and then trace this line around to the back and go a little lower to where the two floating ribs are, this roughly forms the outer boundary of the diaphragm. From this circumference the diaphragm domes up to form a peak almost to where the nipples are in men—a great deal higher than most people think, and certainly not in the abdominal region where so many people point when they speak about the support of the diaphragm in singing. Right on top of the peak of the diaphragm is the heart, and on either side, quite high up and way up into the back and armpits, are the lungs—also a great deal higher than we normally think they are.

Fig. 40. The diaphragm

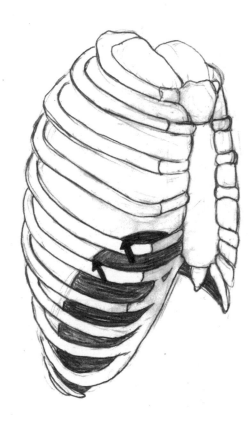

Inhalation

Exhalation

The diaphragm is an extremely active and hard-working muscle, contracting steadily all our lives to ensure a constant supply of air to the lungs. When the diaphragm contracts, the dome moves downward and flattens; this increases the space in the lower thorax and causes air to flow into the lungs (Fig. 40). At the same time, the ribs ascend, which also contributes to the expansion of the space within the chest cavity and the flow of air into the lungs (Figs. 40 and 37). When the ribs return to their lower position and the diaphragm ascends, air rushes out. The active movement of the diaphragm is downward, not upward, as many people think (perhaps because the ribs go up in inhalation); when the ribs *rise,* the diaphragm *descends.* The upward movement of the diaphragm is the passive phase; when the diaphragm relaxes, the ribs descend, and air is forced out of the lungs. When we hold our breath, we are holding the diaphragm down; learning to let go of the breath means to allow the full *ascent* of the diaphragm.

Supporting the upper rib cage are the ***scalene*** muscles (Fig. 43). The scalene muscles originate at the upper cervical vertebrae and attach to the upper ribs; they are considered muscles of respiration because they come into play in forceful inhalation, and because they support the ribs from above. The upper ribs, however, do not actually contribute much to breathing, being quite squat and immobile; the scalene muscles are not so much involved in breathing as they are in supporting and suspending the upper rib cage. We'll discuss the muscles that support the thorax in Chapter 19.

123

18. The Abdominal Muscles

Let's turn now to the muscles of the abdomen (Figs. 41 and 42). If you look at the muscles of the ribs, or upper part of the trunk, and compare them to the muscles of the abdomen, you'll see that just as in the thorax, the muscles in the abdominal region run obliquely and are simply continuations of the muscles of the thorax. In fact, the entire trunk is a large tube criss-crossed with large sheets of muscle that encircle the whole thing. Muscles also run vertically up the front of the abdomen, and horizontally around the circumference of the abdomen. But the point to remember is that the oblique muscles in the thorax and abdomen are really continuous sheets that form a spiral musculature encircling the entire trunk. We'll look at the spiral musculature of the trunk in Chapter 20.

The muscles of the abdomen are comprised of three layers corresponding to the external and internal intercostal muscles and the transversus thoracis, the muscle underlying the sternum. We often think of the abdominal region as the belly, but in fact it extends from the costal arch and the xiphoid process (the bottom pointed section of the sternum) to the pubic bone below. Much of the musculature of the abdominal area—particularly at the center—is aponeurotic tissue, which is flat, tendinous tissue, so that there are oblique sheets of muscle at the sides of the abdomen that terminate in flat sheets of tendinous tissue toward the belly. This tissue comes together to form a vertical tendon running from the pubic bone to the xiphoid process of the sternum. This midline of tendinous material is called the "linea alba," which means "white line."

There are four primary muscles in the abdominal region. The ***external abdominal oblique*** muscle originates at the eight lowest ribs; its fibers run directly downward to attach to the rim of the pelvis, or iliac crest, and

Fig. 41. The abdominal muscles

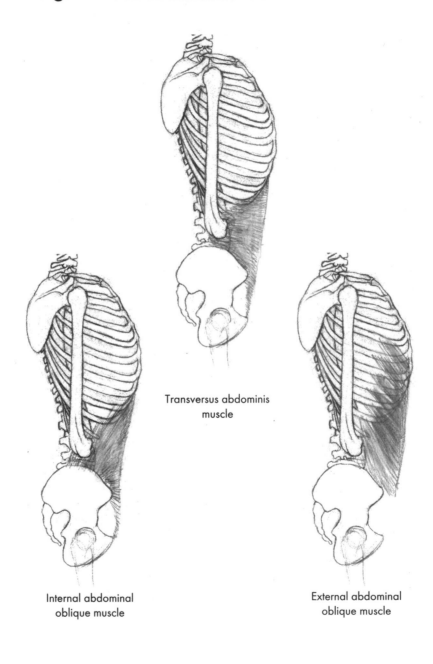

Transversus abdominis
muscle

Internal abdominal
oblique muscle

External abdominal
oblique muscle

obliquely downward and forward to end in aponeurotic tissue that terminates at the linea alba (Fig. 41). This muscle is continuous with the external intercostal muscle of the thorax.

The ***internal abdominal oblique*** muscle originates at the iliac crest and the lumbar fascia (Fig. 41). Its fibers fan out to attach to the pubic bone, the linea alba, the cartilage of the eight, ninth, and tenth ribs, and the lower three ribs. It is continuous with the internal intercostal muscle of the thorax.

The third and deepest muscle, corresponding to the transversus thoracis muscle of the chest, is the ***transversus abdominis,*** or ***transversalis*** (Fig. 41). It originates at the crest of the ilium, the lumbar fascia, and the ribs and runs, as its name suggests, horizontally around the midriff. It terminates in aponeurotic tissue, joining the linea alba and the pubic symphysis. According to *Gray's Anatomy,* the lumbar fascia at its origin can be considered to be its posterior aponeurosis, so that it is basically tendinous tissue at the back, becomes muscle at the sides, and then ends again in tendinous tissue.

Rectus abdominis is the longitudinal muscle popularly known as the "abs" which forms the muscular bumps along the front of the abdomen (Fig. 42). It originates at the pubic symphysis and runs vertically up each side of the linea alba at a slightly outward angle to insert into the fifth, sixth, and seventh ribs; it is enclosed within a sheath of tendon formed by the oblique and transversalis muscles. This muscle is the only one that can be observed on the abdomen, the others being flat sheets that wrap around the abdomen without being prominent. Rectus abdominis is a flexor of the trunk. By acting on the rib cage, this muscle powerfully contracts or flexes the front of the body (as in performing a sit-up). As part of the system of flexors in front of the body which counterbalance the extensors in back, rectus abdominis also serves an important role in helping to maintain erect posture. When the abdominals are not able to contract (this happens if you have had abdominal surgery or a Caesarean section in childbirth), it becomes difficult to walk and even to maintain fully erect posture.

Fig. 42. Rectus abdominis muscle

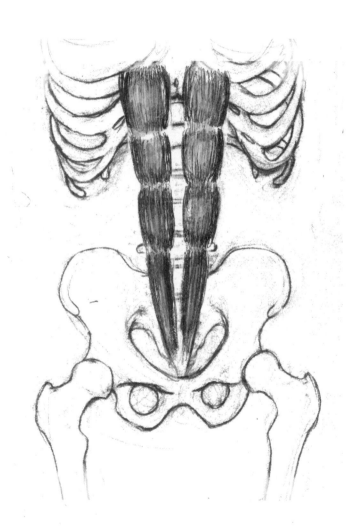

There is a fifth muscle, called ***pyramidalis,*** which lies in front of rectus abdominis at the pubic bone (not pictured); it is a tensor of the linea alba.

External oblique, internal oblique, rectus abdominis, and transversalis form the front and side walls of the abdomen. These walls tend to sag with age, not only because the muscles lose tone, but also because of loss of support in the torso and shortening in the front of the body, which allow the internal organs to sag forward. The posterior wall of the abdominal region is formed by the muscles on the front of the spine—the back or lumbar portion of the diaphragm, iliacus, the psoas muscle, and quadratus lumborum. We'll take another look at these muscles when we discuss the pelvic region.

19. Suspensory Muscles
of the Thorax

Let's turn now to the muscles that maintain the width and support of the ribs and the rib cage. The rib cage as a whole is supported by various muscles. Earlier, we saw that the *scalene* muscles support the upper two ribs from the cervical spine. There are three scalene muscles—*scalene anterior, scalene medius,* and *scalene posterior* (Fig. 43). Scalene anterior and scalene medius originate at the transverse processes of the middle and lower vertebrae of the neck and insert into the first rib. Scalene posterior originates at the lower cervical vertebrae and inserts into the second rib. These muscles are often described as respiratory muscles because they are active in forceful inspiration, but it is more accurate to say that they act as tensile members or guy wires for supporting the rib cage from the upper spine, and in this sense they function more to provide tensile support for the breathing apparatus than to assist in breathing itself.

One of the most important muscles on the front of the body and rib cage is the *sternocleidomastoid muscle* (Fig. 44). It originates by two heads at the sternum and clavicle and attaches to the mastoid process of the skull. This muscle is a flexor of the head and neck, and assists in rotation of the head. Like the scalene muscles, it is considered an accessory muscle in breathing and noticeably contracts during forced inspiration. However, it would be more accurate to say that the sternocleidomastoid muscle provides tensile support for the thorax as a whole, and in this sense is a crucial suspensory muscle for the rib cage and the front of the body.

Notice that, like many of the muscles of the neck and throat, the sternocleidomastoid muscle attaches to the base of the skull. This is highly significant for two reasons. One, it appears at first glance that the rib cage and the flexors on the front of the body are not directly related to the head. In fact,

Fig. 43. The scalene muscles

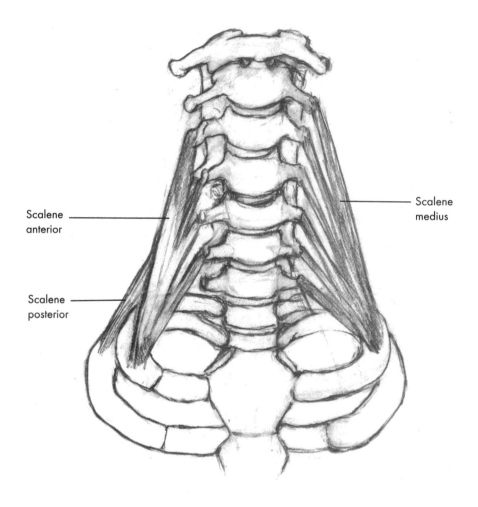

Scalene
anterior

Scalene
posterior

Scalene
medius

however, the rib cage is suspended from the skull by the sternocleidomastoid muscle. This means that, like the extensors at the back of the body, the flexors on the front of the body relate directly to the balance of the head on the spine.

Two, notice that the mastoid process—the point of attachment of the sternocleidomastoid muscle on the skull—is behind the point of support where the head balances on the spine. This means that although a downward pull of the sternocleidomastoid muscle tends to drag the neck forward, it pulls the head *back* at the atlanto-occipital joint. In order, therefore, to lengthen the front of the body and to support the rib cage, the head must release at the back of the neck to lengthen the spine. So the sternocleidomastoid muscle is a key support for the front of the body, and like the extensors in back, it depends for its proper function on the forward balance of the head on a lengthening spine.

At the bottom end of the rib cage, and acting on the ribs from below, *rectus abdominis* forms part of the flexor sheet on the front of the body that supports the rib cage (Fig. 44). It is a powerful flexor of the trunk and assists in expiration, but it also must remain released so that the rib cage and the front of the body maintain their natural length and support. *Quadratus lumborum,* which runs from the iliac crest to the floating ribs, acts upon the rib cage, but in back, not in front (Fig. 45).

Viewed in this way, the rib cage has a kind of supporting mechanism not unlike that of the larynx. Like the larynx, it is supported or suspended from the head; the upward support of these muscles, in turn, is counterbalanced by muscles that run down to the pelvis in front and back. Like the larynx, the collapse of the rib cage can "depress" the postural muscles, and like the larynx, these supporting "extrinsic" muscles are matched by intrinsic muscles that perform the function of the ribs. Unlike the larynx, however, the ribs are themselves part of the supportive structure of the trunk and spine; the tensile members in the front of the body which act as flexors must balance the extensors in the back, providing a balanced support for the entire body.

Fig. 44. Suspensory muscles of the thorax

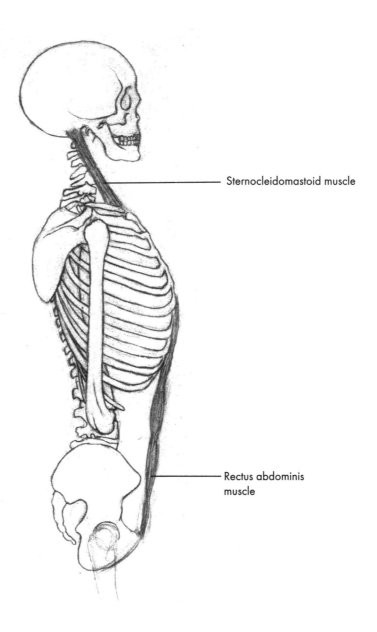

Sternocleidomastoid muscle

Rectus abdominis
muscle

Now let's look at some of the other muscles that support and "widen" the ribs (Fig. 45). We saw that in the region of the floating ribs, ***serratus posterior inferior*** originates at the spinous processes of the upper lumbar and lower thoracic vertebrae, passing obliquely up and outward and breaking into four branches which insert into the four lower ribs. Serratus posterior inferior pulls back and down on the ribs, providing widening and support of the lower rib cage. ***Quadratus lumborum*** originates at the hip bone, or rim of the pelvis, and runs obliquely upward to attach to the transverse processes of the lumbar vertebrae and the lowest rib, which is one of the two free-floating ribs. Quadratus lumborum acts on the floating ribs, helping to provide elasticity and support for the lower back. ***Levatores costarum*** and ***levator costae*** originate at the transverse processes of the vertebrae and, running obliquely downward, attach to the ribs below. These muscles, which we saw earlier when we looked at the deep muscles of the back, are elevators of the ribs. When quadratus lumborum and levatores costarum are functioning properly, the lower back becomes elastic and filled out, the floating ribs move freely, and this lengthening and widening in the back tends to give the entire body increased buoyancy and support.

In the region of the upper thorax, ***serratus posterior superior*** raises the ribs in the upper part of the chest. Finally, ***latissimus dorsi,*** which is the large superficial sheet of muscle covering the entire lower part of the back, helps to maintain the widening and support of the back and thorax (see Fig. 33). Because it attaches to the upper part of the humerus, the widening of the arms helps to maintain the active support of this muscle. When the rib cage is supported properly as a whole and the trunk is lengthening, these "widening" muscles tend to act most efficiently, providing support and elasticity to the action of the ribs and widening the back.

Fig. 45. Muscles of the thorax (cont.)

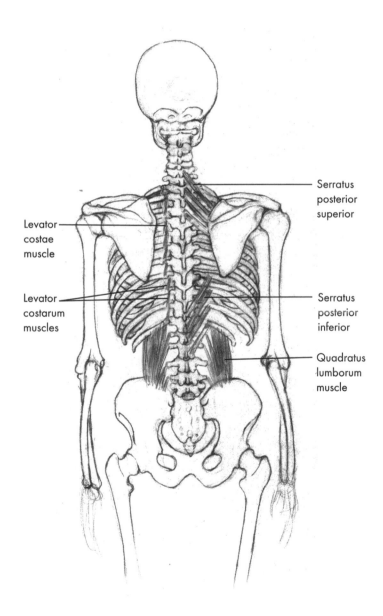

Levator
costae
muscle

Levator
costarum
muscles

Serratus
posterior
superior

Serratus
posterior
inferior

Quadratus
lumborum
muscle

20. The Spiral Musculature of the Trunk

We saw a moment ago that the oblique muscles of the abdomen are continuous with the oblique muscles of the thorax, and that they wrap around the body in criss-crossing spirals that form continuous sheets of muscle encircling the entire trunk, which in this sense can be seen as a cylinder encircled by muscles. In fact, this cylinder includes the head and neck, and the muscular spirals that wrap around it can be traced from the pelvis right up to the head, thus forming a continuous sheet of muscles which extends even into the limbs. From this point of view, the human body can be seen as essentially a tube, beginning at the head and mouth and ending at the pelvis, completely encircled by muscles running around it in a double-spiral or double-helix pattern.

Let's now trace one of these muscular spirals in the trunk, beginning at the right side of the pelvis in front (Fig. 46). We've seen that the internal oblique muscle on the right side of the abdomen originates at the rim of the pelvis and the lumbar fascia, and runs obliquely upward to the midline of the linea alba and the ribs. If you cross the midline and continue this line of direction, the external oblique abdominis muscle on the other side—from its insertion into the midline of the linea alba, the pubic symphysis, and the iliac crest—follows this same line of traction, running obliquely upward and out. That line is continued by the external intercostal which wraps around the body to the levatores costarum of the thorax and the transverse processes of the cervical vertebrae, where it is continued by the oblique deeper muscles of the cervical vertebrae across the transverse processes of the cervical vertebrae and ends at the right occiput.

So, beginning from the pelvis, it is possible to trace a continuous spiral of muscle pulls that begins at the right anterior rim of the pelvis, crosses the

Fig. 46. Spiral musculature of the trunk

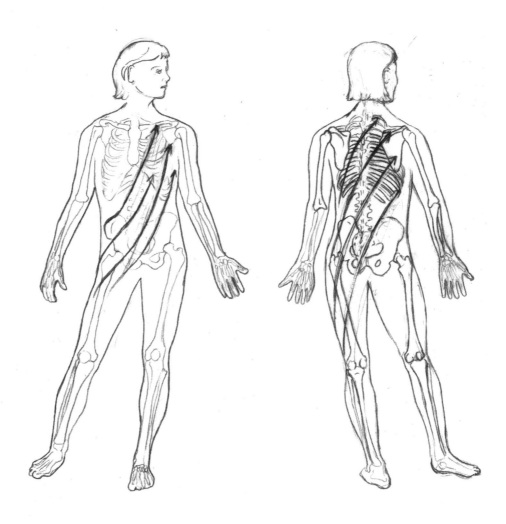

abdomen to the lower left thorax, circles around the ribs to the back, continues obliquely across the transverse processes of the vertebrae across the back, and finally ends at the back of the occiput on the same side as it began. The identical spiral can be traced on the opposite side, so that we end up with two interlacing spirals wrapping around the trunk.

Each of the two spirals can also begin at different points, such as at the back instead of the front, by following the same line of traction but from the different starting point. If you begin the first spiral not at the anterior right rim of the pelvis, but at the back of the pelvis on the left side, then the spiral crosses the ribs in back on the right side, wraps around the ribs and crosses the sternum in front, and ends at the mastoid process of the left side. The second spiral, which begins at the left anterior rim of the pelvis and ends at the occiput of the left side, can also be traced from the right posterior rim of the pelvis to the right mastoid process.

So in one direction you can trace a spiral that begins at the anterior pelvic crest on the right side and crosses to the ribs on the left side, wraps around the ribs to the spine near the neck, and ends up at the occiput on the same side as it started. You can trace the same spiral, but starting at the posterior pelvic crest on the left and crossing the back to the right floating ribs, wrapping around the trunk to cross the sternum, and ending at the mastoid process on the left side. In the opposite direction, one spiral can be traced from the left anterior rim of the pelvis, ending at the left occiput in back. A second spiral can be traced by starting at the right posterior rim of the pelvis and ending at the right mastoid process.

Thus the musculature of the trunk can be viewed not simply as a series of muscles that pull us downward, but as a system of spirals that essentially twist or torque the body asymmetrically. This spiralling musculature contributes to scoliosis and other spinal malformations, but it is present in all people, and particularly in those who are engaged in skilled work which involves dominance on one side and therefore constant asymmetrical twisting of the body musculature. When we slump and collapse in relation to gravity, we don't just pull downward, we twist ourselves posturally. Everyone has a postural twist; it is the key to sorting out repetitive strain injuries that are really complex postural twists. Spirals are also important to understand so that when we look at movement and bodily use, we don't simply move in the sagittal plane, or two-dimensionally, but have a sense of the three-dimensional or spiralling nature of movement.

SHOULDER GIRDLE AND UPPER LIMB

21. The Shoulder Girdle

The shoulder girdle is a yoke-like arrangement of bones suspended above the upper rib cage, providing a mobile structure for supporting the arms and permitting them a wide range of movement (Figs. 47 and 48). It consists of four bones, the two clavicles and the two scapulae, which "float" on top of the rib cage and are not attached to the spine or ribs except where the clavicles join the sternum.

Like the pelvis and legs, the shoulders and arms originally evolved as limbs for supporting and moving the body and are therefore similar in structure. Both the arm and leg are supported by a girdle, or yoke-like framework, which provides a structure for a ball-and-socket joint for the limb; both the arm and leg are comprised of one long bone connecting to two bones, which form a lever system for moving the hand and foot; and the hand and foot are quite similar in structure. In the upright human, however, the arms hang freely and have become modified to be mainly manipulative. The shoulder girdle is attached to the axial skeleton only at the sternum (unlike the pelvic girdle, which is directly and firmly attached to the spine). Also, it is not solid all the way around, which gives the scapula a great deal of freedom. This permits the arm a much greater range of movement than the pelvis does the leg.

The **clavicles** (*clavis,* a key), or collar bones, form the front of the shoulder girdle (Fig. 47). They articulate with the sternum, forming the **sternoclavicular** joint which, again, is the only bony point of contact of the shoulder girdle and arms, which otherwise hang quite freely, with the skeleton. The **scapula** (a Greek word meaning "spade"), which is the wing-like shoulder blade, forms the back part of the "yoke" of the shoulder girdle (Fig. 48). The scapula is a triangular plate of bone; at its upper, lateral corner it forms a kind of spine which culminates in the **acromion process** (meaning "summit of the

Fig. 47. Joints of shoulder girdle

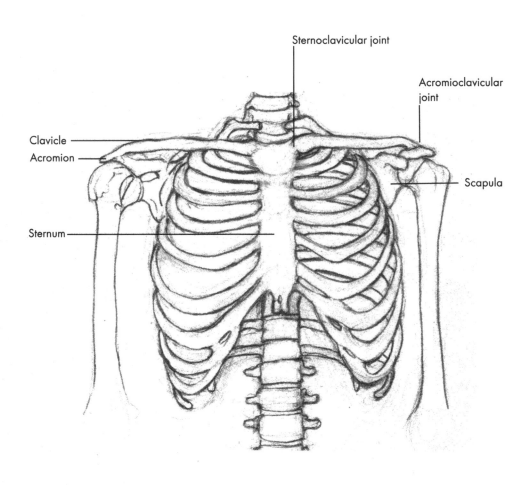

Sternoclavicular joint

Acromioclavicular joint

Clavicle

Acromion

Scapula

Sternum

Fig. 48. Scapula and shoulder joint

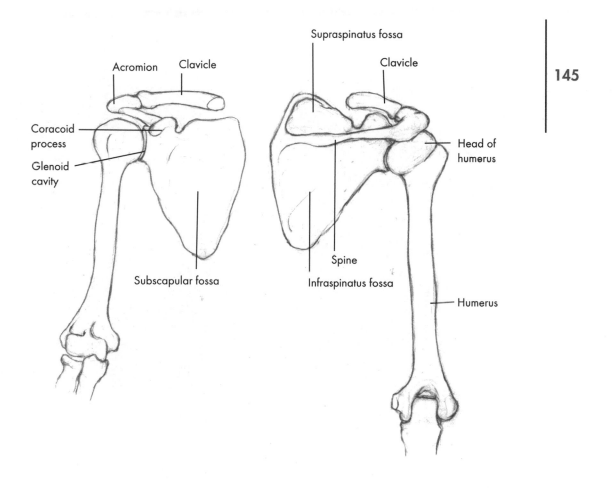

Acromion

Clavicle

Coracoid process

Glenoid cavity

Subscapular fossa

Supraspinatus fossa

Clavicle

Head of humerus

Spine

Infraspinatus fossa

Humerus

Anterior view

Posterior view

Fig. 49. Trapezius, teres major, and latissimus dorsi

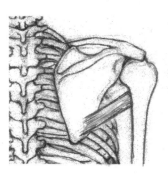

Teres major

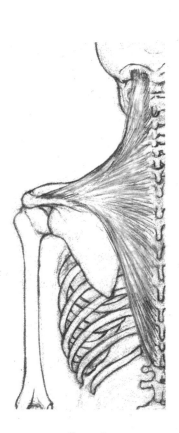

Trapezius

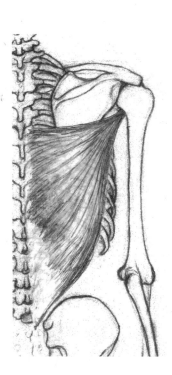

Latissimus dorsi

shoulder"). The outer, or acromial, end of the clavicle forms a small joint with the acromion, called the *acromioclavicular* joint. There is another projection on the anterior side of this part of the scapula called the *coracoid process* (Greek, meaning "a crow") (Fig. 48); it provides an attachment for the pectoral muscle on the front of the chest and for the flexor muscles on the inside of the arm.

Just below the acromion process, the scapula provides a socket, or depression, for the head of the *humerus,* or upper arm, called the *glenoid cavity* (a Greek word meaning "a socket") (Fig. 48). Because the shoulder girdle hangs freely on the rib cage, and because of the shallowness of the glenoid cavity, the arm is permitted a great deal of movement at this joint, which is the most mobile in the body. When we move the arm in a wide-ranging arc, the scapula moves as well, which in turn involves the clavicle, so that when we move the upper arm at the shoulder joint, we involve the entire shoulder girdle. In particular, the scapula is capable of moving along the curve of the ribs and of rotating, which assists the arm in its various movements, giving it as wide a range of movement as possible.

A number of muscles act on the shoulder girdle (Fig. 49). *Trapezius,* which as we saw is the powerful sheet of muscle covering the upper back, arises from the occiput, the ligamentum nuchae, and the spinous processes of the last cervical and all the thoracic vertebrae, and attaches to the scapula and clavicle. *Latissimus dorsi,* which is the large superficial sheet of muscle covering the lower half of the back, fans out from the sacrum and the entire lower spine to converge into the bicipital groove on the upper shaft of the humerus, or arm bone. Together, the trapezius and latissimus dorsi act on the shoulder girdle from virtually the entire length of the spine from head to pelvis, so that the musculature of the shoulder girdle and arms includes not only the muscles directly surrounding the shoulder joint and arms, but the entire trunk.

Although the shoulders appear to be part of the rib cage, the shoulder girdle is in fact suspended by muscles from above, so that it doesn't rest directly

Fig. 50. Scapula muscles

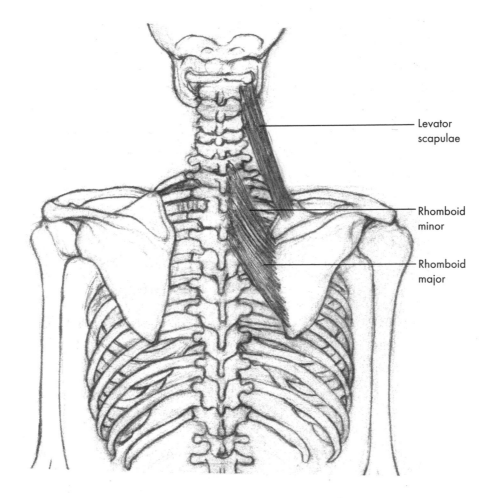

Levator
scapulae

Rhomboid
minor

Rhomboid
major

on the rib cage but "floats" above it. In addition to being supported by muscles, the shoulder girdle is attached, via the clavicles, to the sternum, which of course is indirectly connected to the spine through the rib cage.

Supporting the clavicle, sternum, and upper ribs are the ***sternocleido-mastoid*** and ***scalene*** muscles (Figs. 43 and 44). The sternocleidomastoid muscle runs from the sternum and clavicle to the mastoid process of the skull, suspending the front of the shoulder girdle from the base of the head; the scalene muscles support the upper rib cage and chest.

Completing the shoulder girdle in back, ***rhomboid major*** and ***minor*** originate at the thoracic vertebrae and pass obliquely downward to attach to the vertebral borders of the scapulae. ***Levator scapulae*** originates at the transverse processes of the atlas and upper cervical vertebrae and attaches at the side of the scapula. These muscles help to stabilize and support the scapulae, as well as to elevate it in movements of the shoulder and arm (Fig. 50).

Serratus anterior arises from the eight upper ribs on the side of the body and inserts into the vertebral, or inner, border of the scapula (Fig. 51). This muscle functions mainly to stabilize the scapula when pushing or extending the arms against resistance.

Viewed from this perspective, the shoulder girdle can be seen to be suspended by muscles that support the upper chest, the sternum, the clavicles, and the scapulae from the head, the cervical spine, and the length of the trunk. The shoulder girdle is not meant to collapse or press down upon the rib cage, as so often happens; it functions best when it is suspended lightly on the rib cage, supported by these various muscles and not pinned down to the thorax by a collapse of the torso or by the muscles that narrow across the front of the shoulders.

Let's look now at the muscles that act upon the shoulder in front. ***Pectoralis minor*** (*pectoral,* breast) arises from the third, fourth, and fifth ribs near the sternum and passes upward and outward to insert into the coracoid process of the scapula (Fig. 51). This is a crucial muscle because when it is

149

Fig. 51. Serratus anterior and pectoral muscles

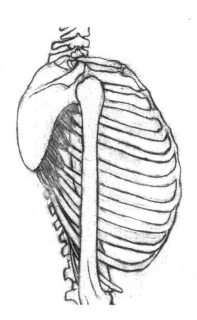

Serratus anterior

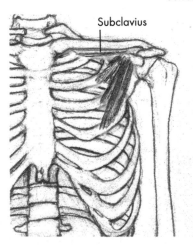

Subclavius

Pectoralis minor

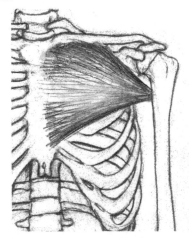

Pectoralis major

chronically shortened, as it so often is in those who use the arms a great deal in desk-work, the shoulders become narrowed in front, with the shoulder joint itself pulled forward. When this happens, the shoulder girdle cannot hang freely and widen, but becomes fixed and shortened in front, which prevents the arms from working properly. Releasing the pectoral muscle so that the coracoid process isn't pinned down and forward allows the shoulder girdle to widen and lighten up off the rib cage.

Pectoralis major is a large, fan-shaped muscle lying on top of pectoralis minor, giving the chest its characteristic shape (Fig. 51). It has a broad origin, extending from the inner half of the clavicle, the sternum and cartilage of the true ribs, and the aponeurosis of the external oblique muscle of the abdomen. From this extensive origin its fibers cross the chest and converge into a tendon that inserts into the bicipital groove on the inside of the upper shaft of the humerus—the same region on the humerus where latissimus dorsi attaches.

Also inserting near the bicipital ridge of the humerus is *teres major,* which originates at the inferior angle of the scapula and runs obliquely upward to insert into the inner part of the upper arm (Fig. 49). It assists in swinging and moving the arm, and is related to the latissimus dorsi muscle.

This area where latissimus dorsi, teres major, and pectoralis major insert into the upper arm (the bicipital ridge on the inside of the humerus) is another crucial region, because when the shoulders are narrowed due to fixation in the pectoralis minor muscle, they are also pulled into the rib cage. When the pull of pectoralis in front and latissimus dorsi in back is released, this reduces the drag of the arms on the rib cage and allows a widening across the upper arms. So, along with the release of pectoralis minor, releasing the shoulder girdle also requires that the muscles attaching to the upper arms release to allow widening of the upper parts of the arms.

Finally, *subclavius* (meaning "under the clavicle") is a small muscle running underneath the clavicle that originates at the first rib and inserts into the underside of the clavicle (Fig. 51). It assists in depressing the shoulder.

22. Muscles of the Arm and Shoulder

Let's turn now to the muscles that support the shoulder joint (known as the rotator cuff muscles because they form a cuff or sleeve over the shoulder joint), and the muscles that act on the shoulder joint. As we saw in the last chapter, the shoulder and arms (in contrast to the hips and legs) are not designed mainly for propulsion, but for manipulative use of the arms and hands in a wide array of activities. For this reason, the shoulder joint is quite shallow, which gives it a great deal of freedom and mobility. However, this also makes the shoulder joint quite unstable; unlike the hip joint, which is firmly bound by very strong ligaments, the shoulder joint has limited ligament support and is largely held in place by the rotator cuff muscles that immediately surround it.

There are four rotator cuff muscles (Fig. 52). *Supraspinatus* (meaning "above the spine of the scapula") originates from the upper area of the scapula above the scapular spine, which it fills up, and inserts right on top of the head of the humerus.

Infraspinatus (meaning "below the spine of the scapula") originates at the large area of the scapula below the spine and converges to insert into the back of the head of the humerus.

Teres minor originates at the outer border of the scapula and attaches to the back of the head of the humerus.

Subscapularis (meaning "under the scapula") is a large, triangular muscle covering the entire under-surface of the scapula. Its fibers pass outward and converge into a tendon that inserts into the front of the head of the humerus.

Passing over the top, front, and back of the shoulder joint, each of these muscles, as well as the *biceps* and *deltoid* muscles, which also pass over the head of the humerus in back and front (and which we'll talk about in a

Fig. 52. The rotator cuff muscles

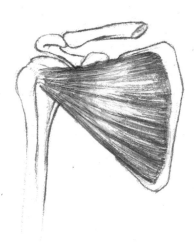

Subscapularis (anterior view)

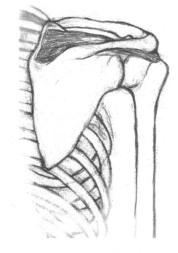

Supraspinatus (posterior view)

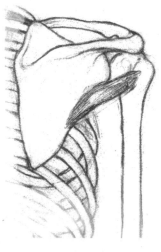

Teres minor (posterior view)

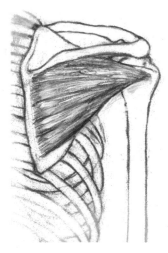

Infraspinatus (posterior view)

moment), assist in maintaining the head of the humerus in the glenoid cavity and moving the arm at the shoulder joint. Because the shoulder joint is supported largely by these rotator cuff muscles, it is much easier to dislocate the shoulder than the hip joint, which is why some young children, whose muscles are still insufficiently developed, tend to dislocate their shoulders when they hang by their arms. It is also these rotator cuff muscles that tend to get torn in sports injuries or falls involving the arms.

The rotator cuff muscles are important to the freedom of the shoulder. In many adults, they tend to become quite fixed; pectoralis minor shortens, causing a narrowing across the front of the shoulders, and the rotator cuff muscles become quite fixed and immobile, leading in extreme cases to the condition known as "frozen shoulder." Along with the release across the shoulders that allows the shoulder girdle to become freely suspended, the rotator cuff muscles must release to restore mobility in the shoulder joint. This in turns allows the scapula muscles to release and to support the arm from the scapula in back, which is where a good deal of the support for the shoulders and arms must come—not from the fixation at the shoulder joint and tightening in the pectoral muscles.

Let's look now at the muscles of the shoulder and upper arm. The ***deltoid muscle,*** which gets its name from its resemblance to the Greek letter Δ, is the large muscle lying on top of the shoulder which gives this area its rounded shape (Fig. 53). It has three sections, the first arising from the outer third of the clavicle, the second from the acromion process, and the third from the spine of the scapula. From these points its fibers converge to extend into the outer side of the shaft of the humerus almost halfway down its length. Its main function is to abduct, or raise, the arm.

On the front of the upper arm, ***coracobrachialis*** is a small muscle which originates at the coracoid process and attaches to the shaft of the humerus (Fig. 54). It assists in elevating and drawing the humerus forward.

Fig. 53. The deltoid muscle

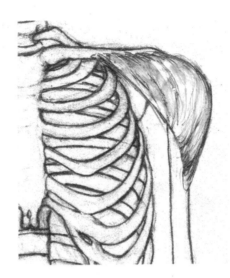

Anterior view

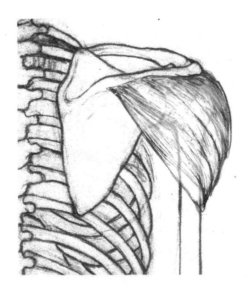

Posterior view

Fig. 54. Flexors of the arm

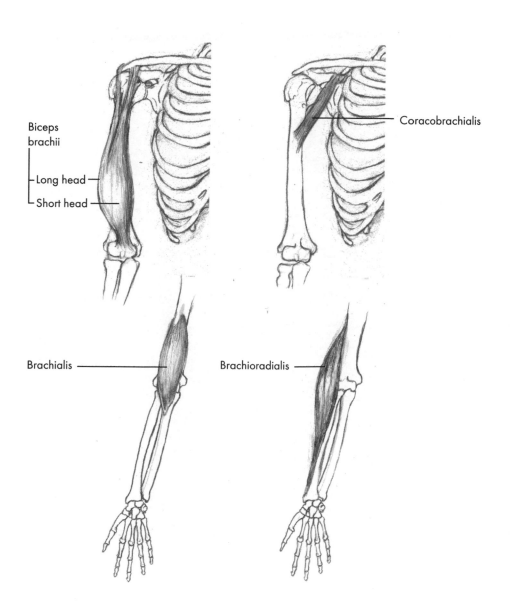

Biceps brachii

Long head

Short head

Coracobrachialis

Brachialis

Brachioradialis

Fig. 55. Triceps brachii muscle

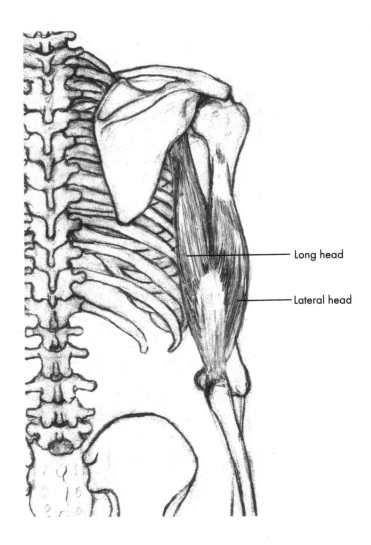

Long head

Lateral head

Posterior view

There are several muscles on the upper arm which flex the arm at the elbow. ***Biceps brachii***, which is the main muscle on the front of the upper arm, has two heads or origins (hence its name "biceps," which means "two heads," and "brachii," which is based on the Latin *brachium,* meaning "arm") (Fig. 54). The short head arises at the coracoid process; the long head originates at the upper margin of the glenoid cavity. The two origins begin as tendons (one of which passes over the head of the humerus), and then become distinct muscle bellies. They converge toward the elbow joint until they again terminate in a tendon which inserts into the radius of the lower arm. You can feel the biceps muscle not only by flexing the arm at the elbow, but also by supinating your forearm when it is flexed at the elbow.

Brachialis and ***brachioradialis*** are also flexors of the elbow (Fig. 54). Brachialis arises from the lower half of the humerus and inserts into the ulna just below the elbow joint. Brachioradialis arises from the lower humerus and attaches to the distal, or far end, of the radius.

On the back of the arm, ***triceps brachii,*** which has three heads, arises from the scapula below the glenoid cavity, from the posterior shaft of the humerus, and from much of the lower part of the humerus (Fig. 55). These three branches converge into a tendon which inserts into the olecranon process, or head of the ulna at the elbow. This muscle is the primary extensor of the arm at the elbow.

23. Muscles of the Forearm

To talk now about the forearm and hand, the forearm consists of two bones, the **ulna** and the **radius.** In anatomical position (palms facing forward), the ulna is the bone that forms the elbow and goes down the inside of the forearm in a line with the small finger. The radius lies on the thumb side of the forearm (Fig. 56). The shaft of the radius and ulna are connected by the **interosseus membrane,** a fibrous sheet that holds them together as one unit. At the end of the radius are eight **carpal** bones forming the wrist, then five **metacarpals,** which are the bones of the hand, and then the **phalanges**—two for the thumb, and three for each finger (see Fig. 58).

We can perform two movements with the forearm. The first is flexion and extension, or bending and extending the arm at the elbow. The two bones that form the elbow joint are the humerus (the upper arm bone) and the ulna, which wraps around the end of the humerus to form the elbow joint (Fig. 56). The end of the humerus forms a rounded groove called the **trochlea,** a Greek word meaning "the sheaf of a pulley," which is exactly what the trochlea looks like. The head of the ulna is shaped like a pincer, which wraps around the trochlea, forming quite a stable hinge joint which is further strengthened by ligaments.

The second movement of the forearm is rotation, or pronation and supination. Just below the elbow joint, there is a notch in the ulna. The proximal head of the radius, which fits into this notch and is held there with a ligament, is able to rotate at this location, forming the **superior radio-ulnar joint** (Fig. 56). You can feel this point just below the lateral epicondyle of the humerus, particularly if you rotate the forearm. At its other end, the distal head of the radius is also able to rotate on the ulna, forming the **inferior radio-ulnar joint** (Fig. 56). When the forearm is rotated, the radius, which lies parallel to the ulna, crosses over the ulna, causing the hand to turn palm

Fig. 56. Bones of elbow and forearm

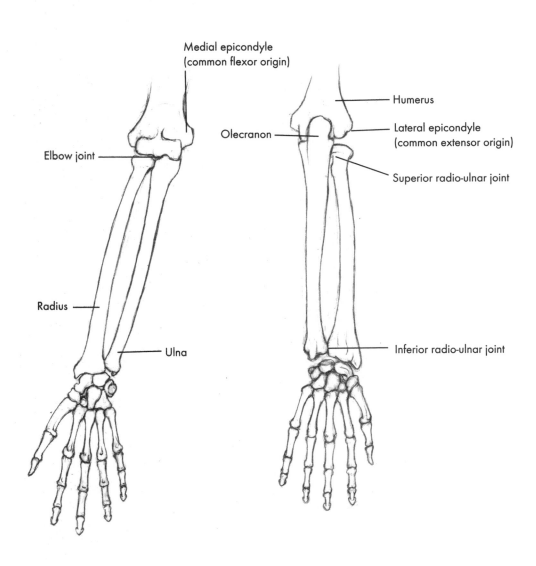

Medial epicondyle
(common flexor origin)

Humerus

Olecranon

Lateral epicondyle
(common extensor origin)

Elbow joint

Superior radio-ulnar joint

Radius

Ulna

Inferior radio-ulnar joint

downward. When it crosses back, the hand is again turned palm upward. The radius, which means "a ray or spoke of a wheel," is given this name because it forms the radius of movement in supination and pronation.

Notice that the bones of the wrist and hand articulate not with the ulna, but with the radius. The purpose of pronation and supination is to be able to bring the hand into position for grasping and manipulation, and because of this, the hand articulates with the radius, the bone that rotates around the ulna. So the ulna is primarily involved in the elbow joint, hinging on the humerus to form the larger lever system for moving the hand toward and away from the body. And the radius is primarily involved in the wrist joint, working together with the hand to rotate around the ulna so that the hand can position itself for grasping and manipulating objects.

A few landmarks around the elbow are worth noting (Fig. 56). The first is the elbow itself, which is formed not by the end of the humerus, but the ulna. It is known as the ***olecranon,*** and it is the weight-bearing point when resting on the elbows or crawling on elbows and knees. It corresponds to the tibial tuberosity, the bony projection below the patellar tendon of the knee, which bears weight when creeping on all fours.

There are two other important landmarks on the elbow: the ***medial epicondyle*** of the humerus (also known as the "funny bone"), which is the bump on the inside of the elbow; and the somewhat less prominent bump opposite the funny bone, which is the ***lateral epicondyle*** of the humerus. These are important points of attachment for the flexors and extensors of the wrist and hand, which we'll look at in the next section.

In the last chapter we looked at the muscles that act upon the forearm to flex and extend the arm at the elbow. ***Triceps brachii*** is the primary extensor or the arm. There are several flexors. ***Biceps brachii*** originates at the coracoid process and the upper margin of the glenoid cavity and inserts into the radius of the lower arm. You can feel the biceps muscle if you supinate your

Fig. 57. Supinators and pronators
of the forearm

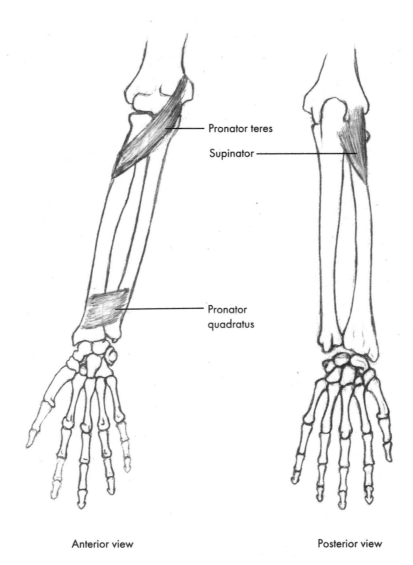

Pronator teres

Supinator

Pronator
quadratus

Anterior view   Posterior view

forearm or flex the arm at the elbow. *Brachialis* arises from the lower half of the humerus and inserts into the ulna just below the elbow joint. *Brachioradialis* originates from the lower humerus and inserts via a long tendon into the distal end of the radius.

There are several pronators and supinators of the forearm (Fig. 57). *Pronator teres,* on the upper forearm, arises from the inner head of the humerus and passes over the anterior side of the radius to attach to its side about halfway down its shaft. *Pronator quadratus* is a square band of muscle just above the wrist attaching the radius to the ulna. These muscles, assisted by the brachioradialus muscle, rotate the radius to turn the palm downward.

Supinator, which has two sections, arises from the outside of the head of the humerus and wraps around the radius to insert into the side of its shaft below the head; it also arises from the posterior ulna just below the elbow and runs across the posterior radius to attach just below its head. Supinator, assisted by the biceps muscle, returns the radius from the pronated to the anatomical position.

The main muscles that act on the hand and fingers are on the forearm, not the hand itself. These fall into two main categories: the extensors, which are on the back of the forearm; and the flexors, which are on the inside of the forearm. These muscles form meaty bundles on the forearm that taper into tendons which attach to the wrist, or which pass over the wrist and attach to the bones of the hand and fingers. There are also several muscles on the hand itself that provide for many of the fine movements of the fingers and thumb; these are the intrinsic muscles of the hand.

The area where the flexor tendons pass over the wrist is quite important for the functioning of the hand. As I said, there are eight carpal bones, and they form an arch on the inner hand that helps give shape to the palm. The carpal bones are held in place on both sides by a ligament, called the *extensor retinaculum* in back and the ***flexor retinaculum*** in front, under which the tendons pass on their way to the hand (see Figs. 61 and 62). The area where the flexor tendons must pass underneath the flexor retinaculum and through the narrow arch formed by the carpal bones is called the ***carpal tunnel;*** "carpal tunnel syndrome" refers to the swelling and pressure that develop in this area from repetitive strain injuries or during pregnancy. Next we'll discuss the wrist and hand and the muscles on the forearm that act on the wrist and hand.

24. Muscles of the Hand and Wrist

In the last section we looked at the forearm and its movements. The pronators and supinator rotate the radius, so that the hand can be positioned facing forward or back. We also saw that the main muscles acting on the wrist and hand are on the forearm, whose muscles work the fingers with tendons that pass through the wrist into the hand through the carpal tunnel. Let's look now at the movements we can make at the wrist and fingers, and the muscles of the forearm that act on the wrist and fingers.

To complete the bones of the wrist and hand, there are eight *carpal* bones which make up the wrist (*karpos,* wrist), arranged in two rows (Fig. 58). In the first row are the *scaphoid* bone (*scaph + oid,* boat-shaped); the *lunate* bone (*luna,* crescent-shaped); the *triquetral* bone (which means "having three corners"); and the *pisiform* bone (which means "pea-shaped"). In the second row are the *trapezium* (Greek, a four-sided table); the *trapezoid;* the *capitate* bone (*caput,* head-shaped); and the *hamate* bone (*hamus,* hook). We commonly think of the wrist as the narrow area of the forearm where we wear a wristwatch, but it is important to keep in mind that the wrist bones make up the heel of the hand and therefore form part of the hand itself. When we crawl on all fours or support weight on the heel of the hand, we are actually bearing weight on the wrist bones. The narrow area of the wrist where we wear our watch is actually where the forearm bones end and the radius articulates with the wrist bones.

It is also interesting to note that two of the wrist bones function as part of the thumb. Unlike the fingers, the thumb starts right at the wrist, and the wrist bones that lie in a line with the thumb—the scaphoid and trapezium—form part of the column of bones which, along with the first metacarpal and the two phalanges, make up the thumb (Fig. 60). The second of these wrist bones—

Fig. 58. Bones of wrist and hand

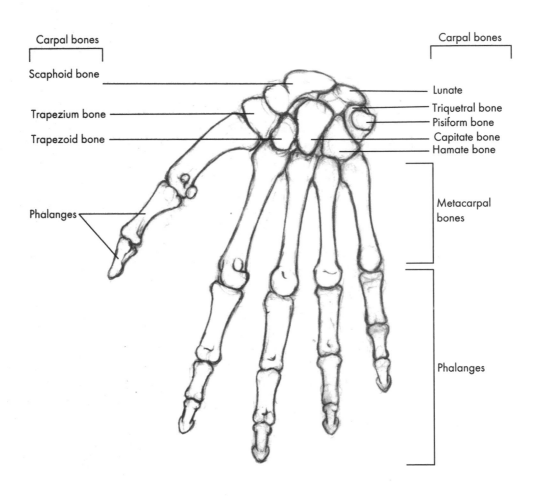

Carpal bones

Scaphoid bone

Trapezium bone

Trapezoid bone

Phalanges

Carpal bones

Lunate

Triquetral bone

Pisiform bone

Capitate bone

Hamate bone

Metacarpal bones

Phalanges

Right hand: palmar view

the trapezium—flexes slightly when you move the thumb, so that the wrist bones actually contribute to the action of the thumb.

As we saw in Chapter 23, the wrist articulates not with the ulna, but with the radius. At its distal end the radius has quite a large surface, and it articulates with the scaphoid and lunate bones to form the wrist joint, called the *radiocarpal joint* (Fig. 59). These two carpal bones form a kind of convex surface like the surface of a football, which corresponds to the concave surface at the end of the radius. At this point, the scaphoid and lunate bones can move in two different ways. They can pivot sideways in relation to the radius, resulting in abduction and adduction of the hand. And they can bend forward and back, resulting in flexion and extension at the wrist. There is another joint in the wrist between the first or proximal row of carpal bones and the second or distal row, called the *mid-carpal joint.* This joint contributes to the ability of the hand to flex at the wrist, and also to abduction and adduction. The main thing to remember about these joints is that we can perform two movements of the hand at the wrist: extension and flexion, and abduction and adduction, or ulnar deviation (moving the hand sideways in relation to the wrist), which take place at the radiocarpal and the mid-carpal joints. In conjunction with rotation of the forearm, which as we saw doesn't take place at the wrist but involves the entire forearm, these movements of the hand at the wrist make it possible to orient the hand in any position for grasping objects.

The bones of the hand are made up of five *metacarpals* which form the palm, and the *phalanges* which make up the fingers and thumb—two for the thumb and three for each finger (Fig. 58). There are several joints in the wrist and hand which are involved in movement of the hand. The hand is capable of various actions such as hollowing the palm (as in grasping or cupping the hand) and opening the palm (as in splaying the fingers). The fingers are also capable of quite complex movements. The last two joints of the fingers, the *interphalangeal* joints, are hinge joints for flexing and extending the fingers.

Fig. 59. Joints of the wrist

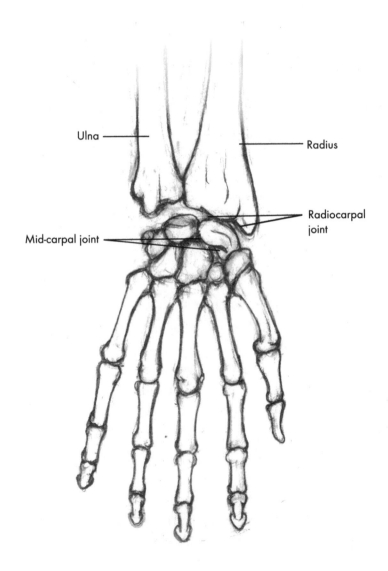

Ulna

Radius

Radiocarpal joint

Mid-carpal joint

The first joint of the fingers, or the knuckle joint, is called the ***metacarpophalangeal*** joint because it is formed by the metacarpal and the phalange. In addition to being able to flex and extend, it is capable of sideways movement, or adduction and abduction.

What makes the hand so extraordinary, of course, is the thumb, which is capable of opposing the fingers and which makes prehension, or grasping, possible. The column of bones which make up the thumb has no fewer than four joints, and there are nine muscles that control its movements. The thumb makes it possible to grasp tools and weapons; it is also capable of the very delicate movements of opposing individual fingers, and dynamic grips such as playing a stringed instrument. Without the thumb, as some anatomists have remarked, the hand loses many of its remarkable functions.

As we saw a moment ago, the thumb is made up of a series of five bones—the scaphoid and trapezium on the wrist, the first metacarpal, and two phalanges—which form a column that starts right at the wrist joint (Fig. 60). The scaphoid and trapezium form the base of this column; the pad of the thumb is formed by the first metacarpal. When we oppose the thumb to the fingers, which of course moves the entire pad of the thumb, we are moving the metacarpal in relation to the trapezium. This joint, called the ***trapezometacarpal joint,*** is a saddle joint and moves quite freely. It is located very close to the wrist and is the crucial articulation that makes opposition of the thumb possible. The ***metacarpophalangeal joint*** of the thumb corresponds to the metacarpophalangeal or knuckle joints of the fingers, although it is much closer to the wrist and larger. Like the metacarpophalangeal joints of the fingers, this joint both flexes and extends, and moves sideways. The ***interphalangeal joint,*** like the corresponding joints of the fingers, simply hinges.

A number of muscles flex the fingers and the hand at the wrist. Many of these arise from a common origin, the ***medial epicondyle*** of the humerus (also known as the "funny bone"), which is therefore called the "common flexor

Fig. 60. Joints of the thumb

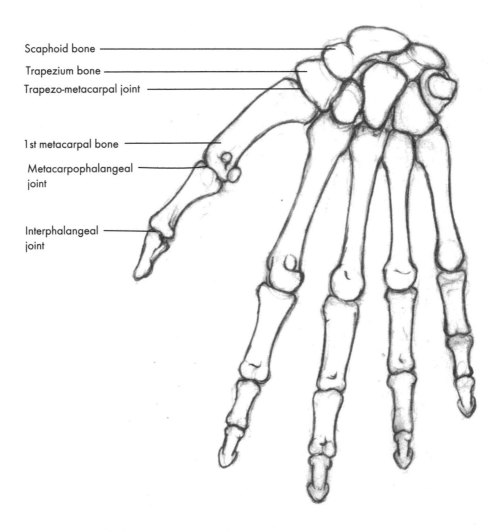

Scaphoid bone

Trapezium bone

Trapezo-metacarpal joint

1st metacarpal bone

Metacarpophalangeal
joint

Interphalangeal
joint

origin"; several extensors arise from the ***lateral epicondyle*** of the humerus, called the "common extensor origin" (see Fig. 56). It is important to remember that the main muscles that act on the hand are on the forearm, not within the hand itself; most of these have long tendons which cross the wrist and work the fingers. The muscles within the hand are responsible for the more delicate movements of the hand and digits.

Let's talk first about the muscles that flex, extend, and deviate the hand at the wrist (Fig. 61). ***Flexor carpi radialis*** arises from the common flexor origin and inserts into the base of the metacarpal bones of the first and second fingers. Its action is to flex and abduct the wrist.

Palmaris longus also arises from the common flexor origin and inserts into the flexor retinaculum and palm. It assists in flexion of the wrist.

Flexor carpi ulnaris arises from the common flexor origin and runs down the ulnar side of the forearm. It inserts into the wrist on the ulnar side; it flexes and adducts the hand at the wrist.

Extensor carpi radialis longus arises from the lateral epicondyle of the humerus and inserts into the base of the metacarpal bone of the index finger. ***Extensor carpi radialis brevis*** arises from the same point and inserts into the base of the metacarpal bone of the third finger. These muscles extend and abduct the wrist.

Extensor carpi ulnaris originates from the lateral epicondyle of the humerus and the ulna and inserts into the base of the metacarpal of the little finger. It extends and adducts the wrist.

Let's look now at the extrinsic muscles of the hand—that is, the muscles on the forearm that act on the fingers. There are two important muscles on the forearm that flex the fingers (Fig. 62). ***Flexor digitorum profundus*** arises from much of the shaft of the ulna and interosseous membrane. It splits into four tendons which pass through the carpal tunnel and insert into the distal phalanges of the four fingers.

Fig. 61. Extensors and flexors of wrist

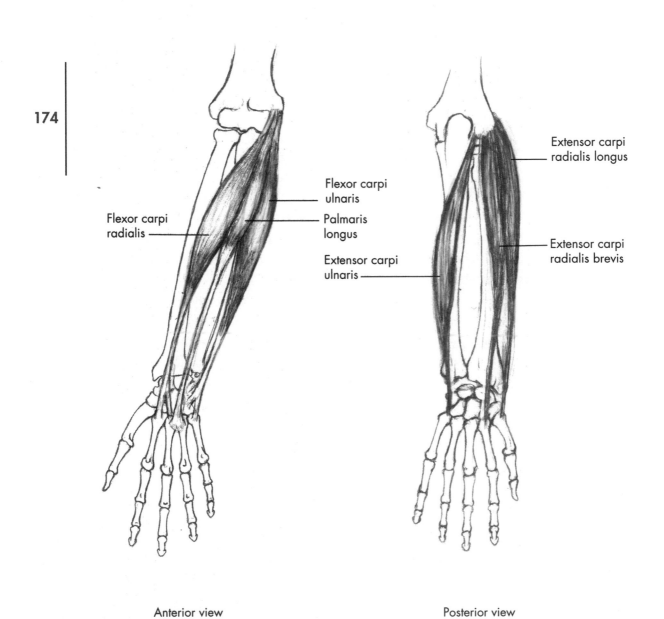

Flexor carpi
radialis

Flexor carpi
ulnaris

Palmaris
longus

Extensor carpi
ulnaris

Extensor carpi
radialis longus

Extensor carpi
radialis brevis

Anterior view

Posterior view

Fig. 62. Flexors of digits

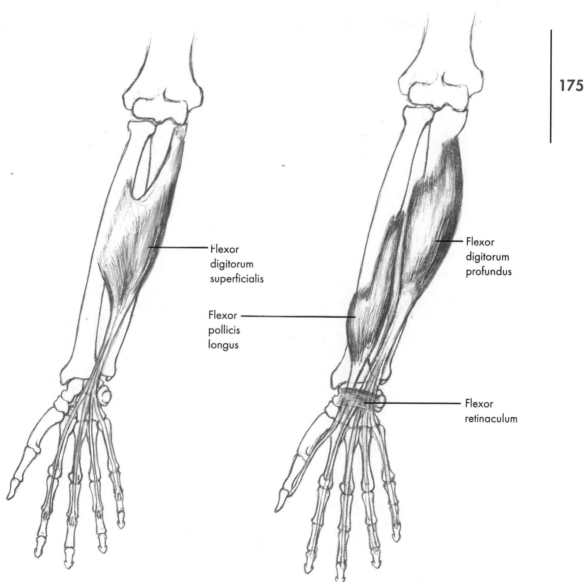

Flexor
digitorum
superficialis

Flexor
pollicis
longus

Flexor
digitorum
profundus

Flexor
retinaculum

Fig. 63. Extensors of digits

Extensors of fingers: posterior view   Extensors of thumb: posterior view

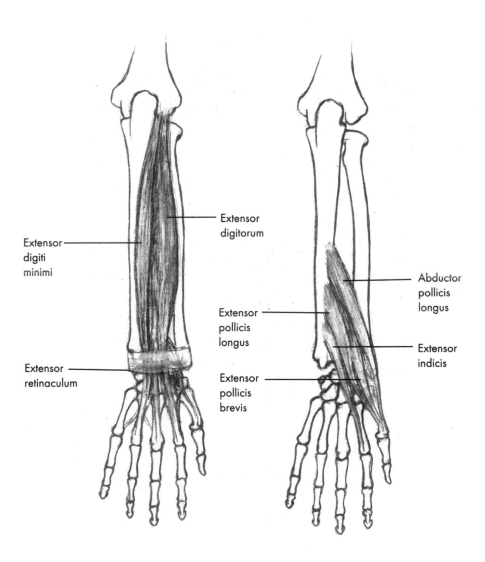

Extensor
digiti
minimi

Extensor
digitorum

Extensor
retinaculum

Extensor
pollicis
longus

Extensor
pollicis
brevis

Abductor
pollicis
longus

Extensor
indicis

Flexor digitorum superficialis arises from the humerus, ulna, and radius; it splits into four tendons which pass through the carpal tunnel and insert into the phalanges of the four fingers.

These two muscles are powerful flexors of the fingers. Flexor digitorum superficialis flexes the fingers up to the second phalanx. Flexor digitorum profundus flexes all the phalanges, including the distal ones.

There is one muscle, *extensor digitorum,* that extends the fingers (Fig. 63). It arises from the common extensor origin (the lateral epicondyle of the humerus) and splits into four tendons which attach at various points on the back of the phalanges.

Four muscles on the forearm act on the thumb. *Flexor pollicis longus* arises from the shaft of the radius and inserts into the base of the last, or distal, phalanx of the thumb (Fig. 62). This muscle flexes the thumb at its joints, as in grasping an object or hanging from a branch.

Abductor pollicis longus arises from the shaft of the ulna and radius and inserts into the base of the metacarpal bone of the thumb (Fig. 63).

Extensor pollicis brevis arises from the shaft of the radius and interosseus membrane and inserts into the base of the first phalanx of the thumb (Fig. 63).

Extensor pollicis longus arises from the shaft of the ulna and interosseous membrane and inserts into the last phalanx of the thumb (Fig. 63). These muscles make it possible to extend the thumb and to draw it backward towards the wrist.

Finally, there is a muscle on the forearm that extends the index finger, and one that extends the little finger (Fig. 63). *Extensor indicis* arises from the shaft of the ulna and the interosseous membrane and joins the tendon of the extensor digitorum muscle that acts on the index finger. It extends the index finger separately from the others, as in pointing; it also assists in extending the wrist.

Extensor digiti minimi arises from the common extensor origin and joins the tendon from extensor digitorum acting on the fifth finger. It extends the little finger and, like extensor indicis, makes it possible to do so separately from the other fingers.

To recap, the flexors of the wrist and fingers are on the inside of the forearm, and the extensors are on the outside. The flexors originate mainly at the inner, or medial, epicondyle of the humerus (the funny bone); if you pronate the forearm, you can feel where this meaty bundle of muscles attaches to the funny bone (you can also feel where the extensors attach to the lateral epicondyle of the humerus). Because the tendency in typing or writing is to over-contract the flexors of the forearm, it is sometimes necessary to think of lengthening from the fingers to the elbow, which maintains length in these muscles.

25. Intrinsic Muscles of the Hand

We looked in the last chapter at the muscles of the forearm that move the hand at the wrist, flex and extend the fingers, and move the thumb. Let's turn now to the intrinsic muscles of the hand (the muscles on the hand itself), which are responsible for the more precise movements of the fingers and thumb.

The intrinsic muscles of the hand can be divided into three groups: the muscles that act on the thumb; the muscles of the little finger; and the muscles of the palm which occupy the spaces in between the bones of the fingers and which act on the fingers. The intrinsic muscles of the thumb form the fleshy pad of the thumb called the *thenar eminence;* the intrinsic muscles of the fifth finger occupy the ulnar side of the hand and form the *hypothenar eminence,* the muscular pad on the side of the hand above the little finger. The muscles of the palm are not fleshy and so are not particularly visible.

There are four intrinsic thumb muscles; these lie on the palm and form the muscular pad of the thumb (Fig. 64). *Adductor pollicis* originates at the wrist and the third metacarpal bone and attaches to the proximal phalanx of the thumb; its function is to adduct the thumb.

Flexor pollicis brevis arises from the trapezium and the flexor retinaculum and attaches to the proximal phalanx; its function is to flex and rotate the thumb.

Opponens pollicis arises from the trapezium and the flexor retinaculum and attaches to the metacarpal bone of the thumb; its function is to oppose the fingers.

Abductor pollicis brevis arises from the flexor retinaculum and the wrist bones and inserts into the proximal phalanx of the thumb; its function is to abduct the thumb.

Fig. 64. Intrinsic muscles of the thumb

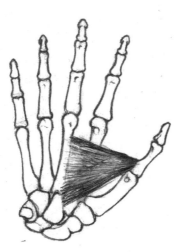

Adductor pollicis

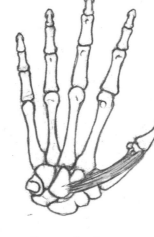

Flexor pollicis brevis

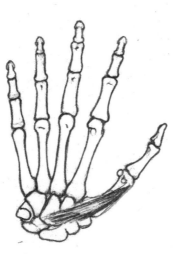

Opponens pollicis

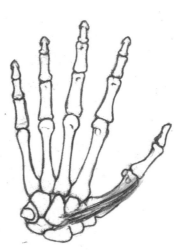

Abductor pollicis brevis

The actions of each of these muscles can be discovered by experimentation. For instance, you will find that the thumb is able to oppose each of the fingers, to press against the side of the index finger, and to rotate in relation to the palm. The movement of the thumb is of course very significant, since it gives us the ability not only to grasp objects and use tools, but to perform all kinds of complex fine motor tasks.

There are three muscles of the little finger, or "digiti minimi," which means "smallest finger" (Fig. 65). They act from the wrist to move the little finger and form the fleshy, muscular area on the ulnar side of the hand. ***Abductor digiti minimi*** arises from the pisiform bone of the wrist and inserts into the base of the proximal phalanx of the little finger. ***Flexor digiti minimi brevis*** originates at the hamate bone and flexor retinaculum just inside of abductor digiti minimi and also attaches to the proximal phalanx of the little finger. ***Opponens digiti minimi*** arises from the hamate bone and flexor retinaculum and inserts into the fifth metacarpal. The actions of these muscles are evident from their names and can be easily discovered by experimentation.

In the middle of the palm, the ***interossei*** (which are so named because they occupy the space in between the metacarpal bones) originate from the metacarpals and insert into the phalanges (Fig. 66). There are four dorsal and three palmar interossei. The dorsal muscles abduct the fingers, or draw them apart; the palmar muscles adduct the fingers, or draw them together.

The ***lumbrical muscles,*** which is Latin for "earthworm," arise from the tendons of the deep flexor muscle of the forearm (flexor digitorum profundus) and insert into the tendons of extensor digitorum (Fig. 66). In conjunction with the interossei, the lumbricales flex the first phalanges and extend the second and third phalanges—that is, flex the fingers at the knuckles but keep them straight from the knuckles, as in holding a brush or grasping a flat object between the fingers and thumb. These muscles make it possible to

Fig. 65. Intrinsic muscles of the little finger

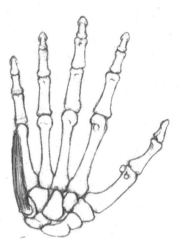

Abductor digiti minimi

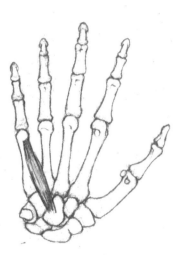

Opponens digiti minimi

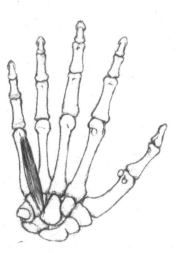

Flexor digiti minimi brevis

Fig. 66. Interossei and lumbricales

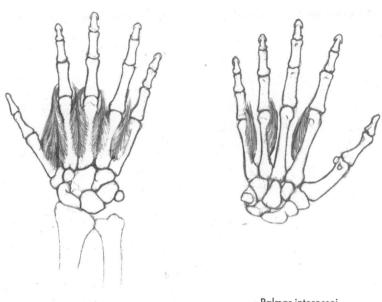

Dorsal interossei   Palmar interossei

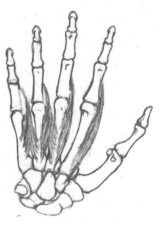

Lumbricales: palmar view

grasp objects with precision and with minimal activity of the larger flexor muscles of the biceps and forearm, which explains why some procedures for learning to use the arms and hands in a coordinated fashion are performed with straight fingers.

PELVIS AND LOWER LIMB

26. The Pelvic Girdle

In contrast to the shoulder girdle, which supports the mobile use of the upper limb for manipulation, the main function of the pelvic girdle is to provide a structure for support and locomotion on two feet. As we saw earlier, the upper and lower limbs are similar in structure. But just as the shoulder girdle provides a mobile framework for limbs that are used not for support but for manipulation, the pelvic girdle has become modified to provide a strong and stable base for support entirely on two legs. Unlike the shoulder, the pelvis is firmly attached to the spine at the sacrum, and the two "wings" of the pelvis are joined in front to provide maximum strength. This also gives the pelvis the capacity to serve as a basin for the internal organs. "Pelvis" is a Latin word meaning "basin."

The pelvis performs two crucial functions. First, it transfers the weight of the body down through the legs into the ground, and it absorbs shock from the legs. Second, it provides a solid structure for the leg muscles, which attach to the pelvis and act upon it to support and move the legs. Because of its weight-bearing responsibility, the pelvis is much more solid and inflexible than the shoulder girdle, capable of absorbing a great deal of shock, and part of a limb system that can exert much more force than the arms. Unlike the shoulder girdle, which is not directly attached to the axial skeleton and which is therefore highly mobile, the pelvis is directly linked with the spine. This sacrifices some of its mobility, but makes it very strong and stable.

The pelvis is structured like an arch that transmits weight from the spinal column into the hip joints and legs. At the lower part of the pelvis are two rocker-like bones, known as the "sit bones"; in sitting, the arch design of the pelvis transmits the weight of the body not into the hip joints, but into these two rockers and onto the sit bones.

Fig. 67. The pelvis, the right innominate bone

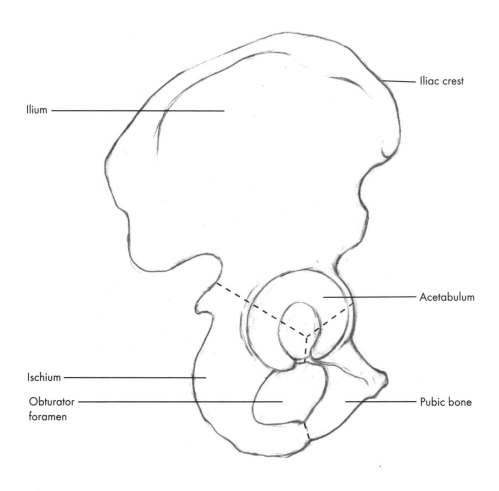

Iliac crest

Ilium

Acetabulum

Ischium

Obturator
foramen

Pubic bone

The two wing-like bones of the pelvis are called the ***innominate*** bones, or ***os innominatum,*** which means "nameless bone" (Fig. 67). Most anatomical structures were named according to what they looked like; having no resemblance to anything, these bones of the pelvis were dubbed "nameless bone"!

The wing-like innominate bones of the pelvis are actually made up of three bones: the ***ilium,*** which means "flank"; the ***ischium,*** which is Greek for "hip"; and the ***pubic*** bones (Fig. 67). The large wing-like part of the bone is the ilium; it forms the prominent hips. The bottom part of the pelvis, which forms the sit bones, is the ischium (the part you actually sit on are the ***ischial tuberosities***) (Fig. 68). The pubic bones form the front part of the pelvis and join the two halves of the pelvis together in front to form a ring-like structure. The three bones of the pelvis—the ilium, the ischium, and the pubic bone— are rigidly attached, so that they function as one whole. This whole is in turn quite firmly attached to the spine where the ilium joins the sacral vertebrae, or sacrum, to form the ***sacroiliac joint*** (Fig. 69). Because of the connection of the pelvis to the sacrum, the pelvis cannot move independently of the skeleton in the way the scapulae of the shoulder can; the pelvis can move only in conjunction with the spine, so that the pelvis and sacrum work as one unit, the sacrum functioning as the back part of the ring of the pelvis. Below the pelvis, the spine has little function, which is why the vertebrae within the sacrum and coccyx, the remnants of the tail, are considered vestigial.

The wings of the pelvis, which correspond to the scapulae of the shoulder, are joined in the back to the sacrum by very strong ligaments, which we'll talk about in a moment. In front the pubic bones form a joint, the ***pubic symphysis*** (Fig. 68), which is cushioned by a disc much like the intervertebral discs. This disc joins the pubic bones together and absorbs shock. In the upright posture, the weight of the trunk is transmitted down through the spine into the sacrum, where the weight is transferred through the sacroiliac joint

Fig. 68. Landmarks of the pelvis

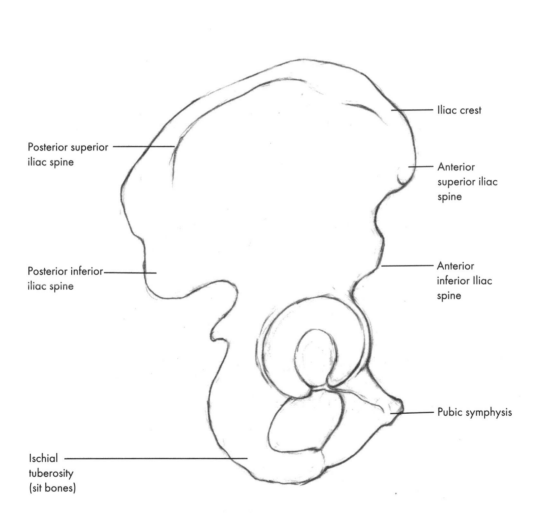

Iliac crest

Posterior superior
iliac spine

Anterior
superior iliac
spine

Posterior inferior
iliac spine

Anterior
inferior Iliac
spine

Pubic symphysis

Ischial
tuberosity
(sit bones)

into the wings of the pelvis, which act as a bridge to then transfer the weight to the hip joints and down through the legs. The wings of the pelvis are connected in front at the pubic bones to form a solid ring, which further stabilizes the pelvis and makes it possible for it to absorb shock through its entire circumference. Unlike the scapulae of the shoulders, the two sides of the pelvis are joined in back (to the sacrum) and in front (at the pubic bones) to form a solid ring. This gives the pelvis tremendous stability and strength. Also, because it is directly linked to the spine, the pelvis is capable of absorbing shock from the legs as well as transmitting weight directly from the trunk into the legs. Because the sacroiliac joint and the pubic symphysis are bound tightly by ligaments, they are capable of little or no movement, except during pregnancy and childbirth, when the ligaments loosen up and allow a slight amount of movement.

There are several points on the pelvis that are important to know (Fig. 68). First, the ilium, or hip bone, forms a bony rim called the *iliac crest* around the waist. At the front, this rim is called the ***anterior superior iliac spine;*** it is what many people incorrectly point to when they are asked to identify the hip joint. The back part of this rim, the ***posterior superior iliac spine,*** along with the sacrum in between, forms the bony part of the back just above the buttocks and below the small of the back. When we lie in the semi-supine or constructive rest position, the weight of the lower trunk rests on the posterior iliac crests (and the sacrum), while the weight of the upper trunk tends to rest on the spines of the scapulae.

The ***hip joints*** are another important location (Fig. 69). If you ask most people to point to their hip joints, they will point to the hip bones—the iliac crests—which of course are part of the pelvis, not the hip joints. In fact, the hip joints are several inches below the iliac crest, about a hands-width apart across the lower pelvis. So just as we locate the head-neck joint too low, we locate the hip joint too high—in other words, we conceive of the trunk as

Fig. 69. The hip joint and femur

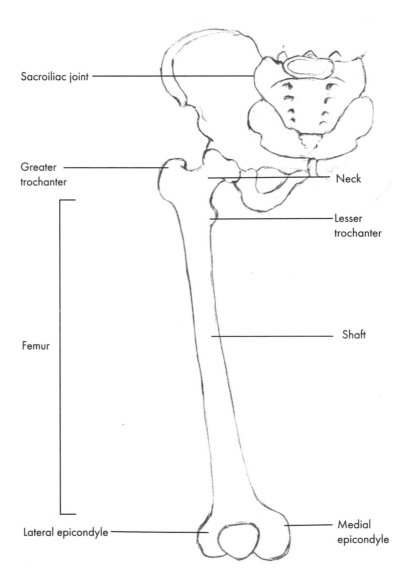

Sacroiliac joint

Greater trochanter

Neck

Lesser trochanter

Femur

Shaft

Lateral epicondyle

Medial epicondyle

much shorter than it actually is, corresponding to our tendency to actually shorten it!

Third, it is important to know where the sit bones are. When sitting, some people slump so badly that they end up resting not on the sit bones, but on the coccyx and even the sacrum, which are not meant to bear weight in this way. If, in contrast, you are in good sitting balance, you can feel the weight resting on two bony points at the bottom of the pelvis: these are the sit bones or *ischial tuberosities* (Fig. 68). They are the two rocker-like projections at the bottom of the pelvis formed by the ischial bones, and we balance our weight on them in sitting just as we balance our weight on our feet in standing.

Finally, the pubic bones form an important point of attachment for muscles and ligaments. The pubic symphysis—the point where the pubic bones come together in front—is the bony area just above the genitalia. Many people think the abdomen ends just below the navel, but the abdominal region extends right down to the pubic symphysis of the pelvis, which defines the lower border of the abdominal region.

The innominate bones, the wing-like bones of the pelvis which correspond to the scapulae of the shoulder, provide a solid socket for the hip joints. This socket is called the *acetabulum,* which referred, in Roman times, to a little cup used at the table for holding vinegar (Fig. 67). The head of the femur, which is shaped like a ball, fits quite snugly into the acetabulum, forming a very strong joint which is nevertheless capable of moving in every direction because it is a ball-and-socket joint. Around the rim of the hip joint is a fibrocartilaginous disc, called the acetabular labrum, which helps to hold the head of the femur in place and deepens the socket. The socket itself, like most synovial joints, is also lined with cartilage.

The *femur,* of course, is the long bone of the thigh (and the longest bone in the body) (Fig. 69). The femur is capable of moving quite freely at the hip joint in every direction. But because of the solidity of the pelvis, the depth of

Fig. 70. Ligaments of the pelvis

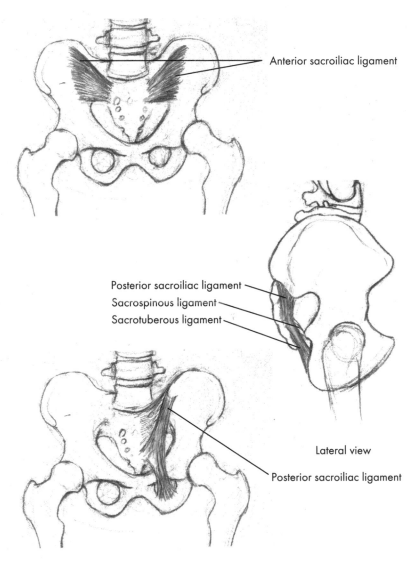

Anterior sacroiliac ligament

Posterior sacroiliac ligament
Sacrospinous ligament
Sacrotuberous ligament

Lateral view

Posterior sacroiliac ligament

Posterior view

the acetabulum, and the rigid attachment of the pelvis to the spine, the legs are not as mobile as the arms, which sit in a much more shallow joint and which are assisted in their range of motion by freely moveable scapulae. If you look at the femur, however, you can see that it is not straight, like the humerus, but has an almost right-angle bend up near the hip joint. This angle increases the range of motion of the femur in the hip joint and, by transferring weight from the hips out to the sides, broadens the base of support of the legs. The longest part of the femur is called the "shaft"; the short portion near the hip joint is called the "neck." The large projection at the bend in the femur, which provides an attachment for the leg and hip muscles, is called the ***greater trochanter*** (which means "to run or roll" in Greek). The projection on the inside of the neck of the femur is called the ***lesser trochanter.*** It is interesting to note that when elderly people break their hips, it is usually the neck of the femur that fractures.

Both the pelvis and the hip joint are firmly bound by a number of ligaments. To look first at the pelvis and spine, we saw a moment ago that the weight of the body is transferred through the sacroiliac joint into the pelvis. This is a crucial area, since the spine is concave at the lumbar region, the sacrum is tilted forward, and there is a great deal of stress at this point, which is sometimes called the ***lumbosacral joint*** (see Fig. 26). The iliolumbar ligaments, which connect the transverse processes of L4 and L5 to the iliac crest, reinforce this region.

The sacroiliac joint is reinforced by several ligaments (Fig. 70). The ***interosseus sacroiliac ligament*** joins the articular surfaces of the sacroiliac joint. The ***anterior sacroiliac ligament*** binds the front of the sacrum to the ilium on each side. The ***posterior sacroiliac ligament*** binds the back of the sacrum to the posterior iliac spine, filling in the space between the sacrum and the protruding crests of the posterior iliac spine. These ligaments not only strengthen the sacroiliac joint, but also help to stabilize movements of the pelvis in relation to the spine.

Fig. 71. Ligaments of the hip joint

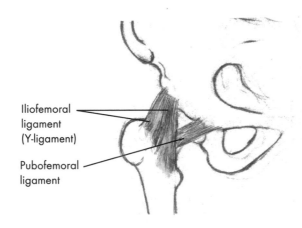

Iliofemoral
ligament
(Y-ligament)

Pubofemoral
ligament

Anterior view

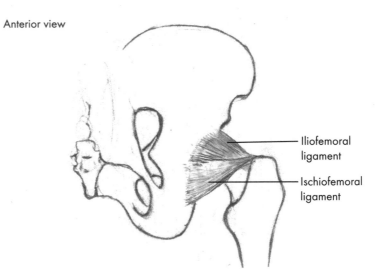

Iliofemoral
ligament

Ischiofemoral
ligament

Posterior view

Lower in the pelvis, the ***sacrospinous ligament*** binds the sacrum to the spines of the ischium, and the ***sacrotuberous ligament*** binds the inside of the sacrum to the ischial tuberosities, strengthening the pelvic basin.

The hip joint has a number of supporting ligaments that reinforce the capsular ligament encasing the joint from the neck of the femur to the pelvis (Fig. 71). The capsular ligament, like other similar ligaments in the body, is lined with a membrane that secretes synovial fluid for lubrication. On the front of the hip joint, the ***iliofemoral ligament*** runs from the anterior inferior iliac spine to the line between the greater and lesser trochanter (called the *intertrochanteric line*). This ligament is sometimes called the ***"Y-ligament"*** because of its shape; it reinforces the front of the hip joint and is extremely strong. The ***pubofemoral ligament*** runs from the lump of the pubic bone called the ***iliopubic eminence*** to the femur. On the back of the hip joint, the ***ischiofemoral ligament*** runs from the ischium, crosses the back of the hip joint, and inserts into the inner surface of the greater trochanter. In the next chapter, we'll look at the muscles of the pelvis and hip.

27. Muscles of the Pelvis and Hip

The pelvis plays a crucial role in posture and movement. Being the main support structure for the legs, it must be able to bear a great deal of stress during the various activities of balancing on two feet, walking, running, bending, and so on. Because of this, both the pelvis and hip joints must be stabilized—a function which is partly served by the supporting, inelastic ligaments and partly by the muscles, which can both lengthen as well as exert contractual power during various movements.

In the earlier chapters on the back and trunk, we looked at three muscles lying on the anterior spine and pelvis that support the pelvis in its relation to the spine (Fig. 72). ***Psoas major*** originates at the lumbar and last thoracic vertebrae and, narrowing as it passes down across the front of the pelvis, inserts into the lesser trochanter of the femur. It corresponds to pectoralis major because, like pectoralis, it inserts into the inner long bone of the limb.

Psoas minor, which is narrower than psoas major and runs on top of it, arises from the sides of the bodies of the lowest thoracic and upper lumbar vertebrae and converges into a tendon which inserts like a strap into the fascia of the ilium, or hip bone, of the pelvis.

Iliacus arises from the inner surface of the wing of the ilium (known as the *iliac fossa*) and passes down across the front of the pelvis to merge with the psoas tendon to insert into the lesser trochanter of the femur. Because the psoas and iliacus muscles are so closely related, they are often referred to as the ***iliopsoas*** muscle.

The function of iliopsoas is to maintain the balance of the pelvis in relation to the spine. When it contracts, iliopsoas rotates the pelvis and pulls the back in, increasing the lordosis or curvature of the lumbar spine. Acting in concert with the extensor muscles of the back, however, iliopsoas balances the action of the extensor muscles lying on the back of the spine. This has the

Fig. 72. The iliopsoas muscle

200

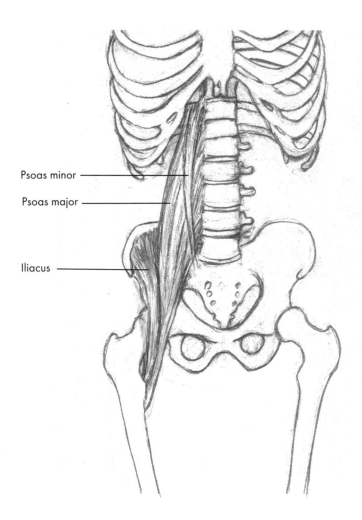

Psoas minor ——————

Psoas major ——————

Iliacus ——————

effect of stabilizing the pelvis and actually serves to straighten, or lengthen, the lumbar spine.

The muscles lying on the front of the lower spine roughly correspond to the pectoral muscles at the front of the shoulders, which as we've seen tend to cause a narrowing across the shoulders. In the same way, the iliopsoas muscles and the hip flexors tend to shorten and tighten the region in front of the hip joints and must release in order to allow freedom in the hips and lengthening of the spine.

The floor of the pelvis is made up of three muscles which form a diaphragm or wall separating the urogenital region or perineum from the pelvic contents above (Fig. 73). The ***levator ani*** muscle runs across the bottom of the pelvis from the pubic bones to the ischium. It contains an opening for the anus and is involved in defecation. Running across the back of the pelvis are the ***piriformis*** and the ***coccygeus*** muscles. Two of these muscles, ***levator ani*** and ***coccygeus,*** lie completely within the pelvis and form the pelvic diaphragm. Although these muscles are not involved in overt movement or support, they are nevertheless important to keep in mind because of the tendency to constrict in this area, which interferes with the freedom of the hip joints and the pelvic region as a whole.

There are six deep muscles of the hip joint, all originating on the pelvis and inserting into the region of the greater trochanter of the hip (Fig. 74). They roughly correspond to the rotator cuff muscles of the shoulder in supporting the hip joints.

Piriformis (which means "having the shape of a pear") originates at the front of the sacrum within the pelvis and inserts on the top of the greater trochanter of the femur.

Quadratus femoris is a short, quadrilateral muscle ("quadratus" means "square") that arises from the lateral part of the rocker forming the tuberosity of the ischium and inserts into the back of the femur between the greater and lesser trochanters.

Fig. 73. The pelvic diaphragm

202

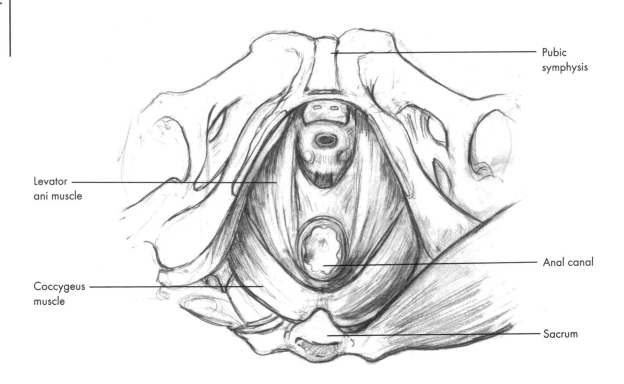

Pubic symphysis

Levator ani muscle

Coccygeus muscle

Anal canal

Sacrum

Fig. 74. The deep muscles of the hip

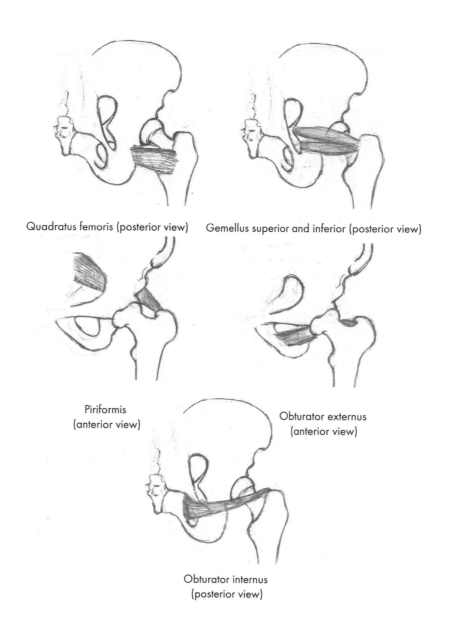

Quadratus femoris (posterior view)

Gemellus superior and inferior (posterior view)

Piriformis
(anterior view)

Obturator externus
(anterior view)

Obturator internus
(posterior view)

Fig. 75. The gluteals

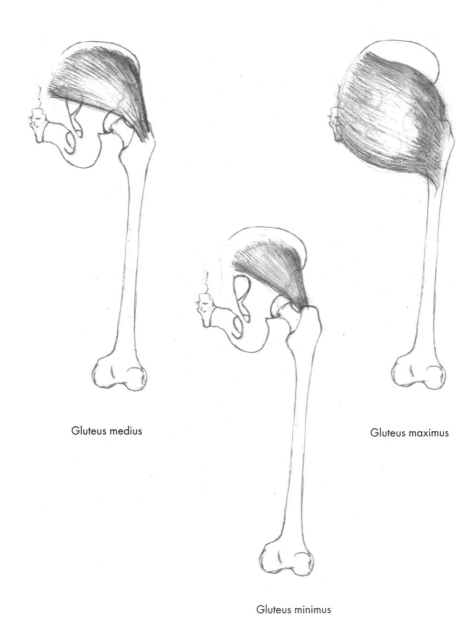

Gluteus medius

Gluteus maximus

Gluteus minimus

Obturator internus arises from the obturator membrane on the inside of the pelvis (which covers the obturator foramen, the holes in the pelvis formed by the ischium) and the bones adjacent to it. It passes through the sciatic notch, wraps around the ischium, and attaches to the lesser trochanter.

Obturator externus arises from the external surface of the obturator membrane and the bony surfaces below its margin, passes across the back of the hip joint, and inserts into the neck of the femur at the base of the greater trochanter.

The *gemelli,* which is based on the word "gemini," meaning "twins," are a small pair of muscles supporting obturator internus. *Gemellus superior* originates from the spine of the ischium and passes horizontally to join the tendon of obturator internus. *Gemellus inferior* originates at the upper part of the tuberosity of the ischium and passes outward to join the tendon of obturator internus.

The obturators, which in Latin means "to stop up" (as in stopping up an opening), are given this name because they cover up the obturator foramen. Together with the gemelli muscles, the obturators form a "hammock" supporting the hip joints. Along with piriformis, these muscles pull the pelvis forward, thus balancing the action of iliopsoas. Together these muscles create a balanced relation of the hip joints to the spine.

The *gluteal* or buttocks muscles (*"gloutos"* is Greek for "rump") are the most superficial muscles of the hip region (Fig. 75). Like the six muscles of the hip joint, they also lie on the posterior hip region and so are really muscles of the hip joint as well as part of the system of extensor muscles lying on the back of the body.

Gluteus medius is fan-shaped, arising from the outer surface of the ilium below the iliac crest and converging to insert into the greater trochanter.

Gluteus minimus is fan-shaped like gluteus medius, but smaller; it originates on the ilium just below gluteus medius and converges to insert into the front of the greater trochanter.

Gluteus maximus, which corresponds to the deltoid muscle of the shoulder, is the largest of the three gluteal muscles and also the bulkiest muscle in the body, giving shape and prominence to the buttocks. It originates broadly at the back of the pelvis from the ilium, the sacrum, and the side of the coccyx and, passing obliquely downward, inserts with tensor fascia latae into the band of tissue running down the outside of the thigh, or *iliotibial tract*, and the upper shaft of the femur.

The three gluteal muscles are extensors of the hip. Gluteus minimus and medius abduct and rotate the thigh medially and help to stabilize the pelvis sideways in walking. Gluteus maximus is the main extensor of the hip. Its primary function is to maintain the trunk upright on the hip joints.

The muscles of the hip also play an important role in sciatica. The sciatic nerve, which is a large nerve in the leg, passes through a notch in the obturator foramen and in between piriformis and quadratus, which can spasm and squeeze the nerve. Although sciatic pain can be caused by disc trouble in the lumbar region, in many cases such pain is lessened by relieving the spasm in these buttock muscles, which means that muscular tension, and not disc trouble, may be the real cause. It is also important to keep in mind that "sciatica" has a number of causes and has come to be used as a blanket term describing any referred nerve pain down the buttocks and leg, even when it does not specifically relate to the sciatic nerve.

28. Muscles of the Thigh

In the last chapter we looked at the muscles of the hip area, which as we saw are mainly supportive in function, roughly corresponding to the rotator cuff and pectoralis muscles of the shoulder girdle. These hip muscles, which tend to be short and squat in design, function mainly to stabilize the pelvis for postural support and against the constant pressures exerted upon it during walking and other activities. In contrast, the muscles of the upper leg tend to be long because they are designed not simply for support but to produce the fluid, large-scale movements of walking and running. It is also interesting to note that there are many more muscles acting on the hip joints and upper leg than on the shoulder and arms.

There are three main groups of long muscles in the leg: the flexors, or *hamstrings*, which are on the back of the legs; the ***adductors*** on the inside of the thigh; and the extensors, or ***quadriceps*** group, which are on the front of the thigh.

To begin with the adductors, if you look at a skeleton, the femur does not drop directly down from the hip joint, but actually comes out from the joint at an angle and then makes a sharp angle downward. It then drops, not directly down, but angles inward, leaving a triangular space in the inner thigh. The adductors fill in this space in the inner thigh (Fig. 76).

Pectineus is a flat, quadrangular muscle which runs from the side of the pubic bone to the inner shaft of the femur just below the lesser trochanter and above the linea aspera, a ridge along the shaft of the femur which serves as a point of attachment for muscles. ***Adductor brevis*** is somewhat triangular in shape and runs from the pubic bone just behind gracilis, inserting into the upper shaft of the femur just below pectineus.

Fig. 76. The adductors

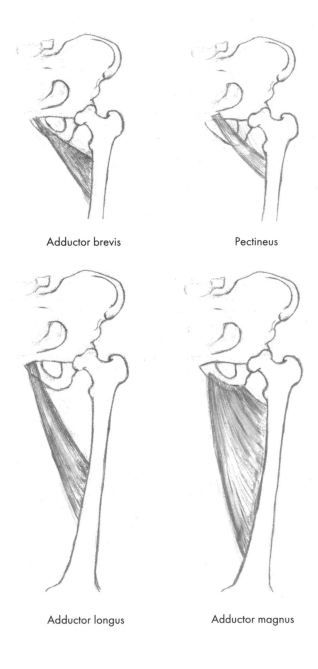

Adductor brevis

Pectineus

Adductor longus

Adductor magnus

Adductor longus, which is also triangular, attaches at the pubic bone just below the pubic symphysis and fans out to insert into the linea aspera toward the mid-thigh.

Adductor magnus is the largest of the adductor muscles. It originates at the pubic bone and the bar, or ramus, of the ischium, fanning out to attach broadly to practically the entire length of the femur from the lesser trochanter right down to the inner, or medial, femoral condyle.

Gracilis, which means slender or thin, originates just below the pubic symphysis and, passing down the inner thigh, inserts into the inner surface of the shaft of the tibia just above the tibial tuberosity (Fig. 77). It is considered an adductor because of its location on the inner thigh, but it is a two-joint muscle (meaning that it crosses two joints) and relates mainly to large movements of the thigh. Pectineus and the three adductor muscles, however, are true adductors, and together they act on the inner shaft of the femur along its length.

The adductors function to stabilize the leg sideways when weight is on it, as in walking. They are also very important in stabilizing the pelvis and assisting in both flexion and extension of the legs, although they seem to be most active in extension. The adductors tend to become quite shortened and contribute to shortening in stature; when they release, this allows the knees to go forward and away to lengthen the spine. To adduct means "to bring toward the mid-line;" it is based on the Latin *ad + ducere,* which means "to lead toward."

The second group of muscles, the extensors, are on the front of the thigh. *Sartorius,* the longest muscle in the body, is thin and ribbon-like (Fig. 77). It arises from the anterior iliac crest and, obliquely crossing the front of the thigh, inserts into the upper part of the inner surface of the tibia. "Sartorius" means "a tailor"; the muscle was so named because of its role in rotating the leg to sit in the tailor's cross-legged position.

Fig. 77. Muscles of the thigh

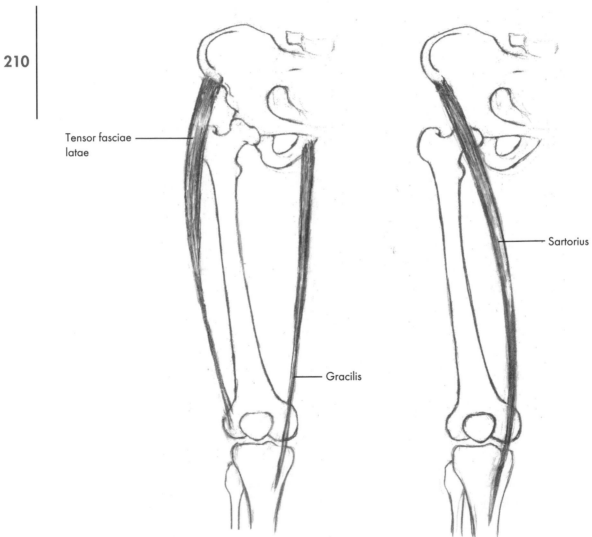

Tensor fasciae
latae

Gracilis

Sartorius

Tensor fasciae latae originates from the anterior iliac crest, or hip bone, and inserts into the ***iliotibial tract,*** which is a band of strong fibrous tissue running down the outside of the thigh that crosses the knee and inserts into the tibia (Fig. 77). Along with gluteus maximus, which as we saw also inserts into the iliotibial tract, this muscle helps to stabilize the hip and knee in standing and walking.

Quadriceps, which is the powerful muscle on the front of the thigh, is the main extensor of the leg at the knee (Fig. 78). It is really made up of four muscles ("quadriceps" means "having four heads"), all attaching to the same point above the knee. ***Rectus femoris,*** so named because of its straight course down the thigh ("rectus" means "straight"), originates at the ilium in two places (from the anterior inferior iliac spine and from just above the acetabulum) and runs down vertically to the knee, where it inserts by a flattened tendon into the patella, which then attaches by a tendon into the tibial tuberosity. This patellar tendon is called the "common tendon" because it is a common point of insertion for all the quadriceps muscles.

Vastus lateralis and ***vastus medialis,*** which mean "great lateral muscle" and "great medial muscle," arise from the sides and back of the upper femur. Their fibers wrap obliquely around the femur on each side to meet in front and converge into the common tendon.

Vastus intermedius arises from the upper two-thirds of the femoral shaft and converges into the common tendon.

The quadriceps muscles extend the leg powerfully at the knee and are also part of the extensor system that supports us against gravity. Some of the extensors at the knee, such as rectus femoris, are also flexors of the leg at the hip.

The third group of muscles is the hamstrings (Fig. 79). The hamstrings flex the leg at the knee and extend the leg at the hip. When working in concert with the calf muscles, they can also act as extensors of the leg at the knee.

Fig. 78. The quadriceps muscles

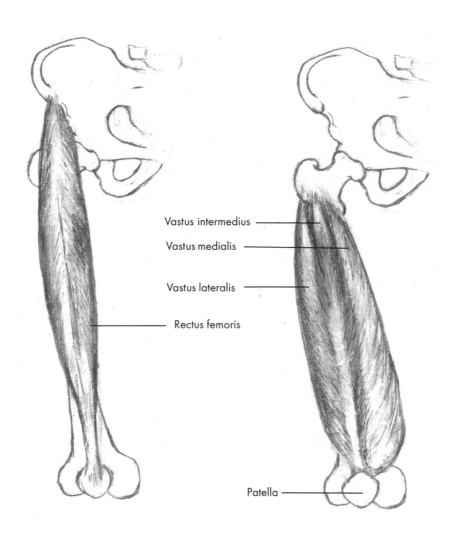

Vastus intermedius

Vastus medialis

Vastus lateralis

Rectus femoris

Patella

The word "hamstring" comes from the practice of cutting these muscles in domesticated animals so that they are "hamstrung"—that is, incapable of flight. *Biceps femoris,* as its name implies, has two heads. One of these originates at the ischial tuberosity, and the other at the lower half of the shaft of the femur. The two muscles merge and insert into the head of the fibula.

Semitendinosus arises at the ischial tuberosity and is inserted by a long tendon into the inner surface of the upper shaft of the tibia.

Semimembranosus also arises at the ischial tuberosity and inserts into the medial condyle of the tibia.

All three hamstring muscles originate from the sit bones; when we sit on the edge of a chair, we can feel the chair press into these muscles just in front of the sit bones. The hamstring muscles are responsible for the tightness in the back of the legs which makes it difficult for many people to sit on the floor with their legs extended in front of them, or to touch their toes. All three hamstring muscles cross the back of the knee and insert into the lower leg bones—two into the tibia and one into the fibula. You can feel the two tendons that insert into the tibia on the back of the knee on the inside, and the tendon that inserts into the fibula on the back of the knee on the outside.

Having now discussed the muscles of the thigh according to their location (hamstrings on the back, adductors on the inner thigh, and extensors on the front), let's look at their role in large-scale movements of the legs. We saw that the short muscles of the hip function mainly to move the leg and to support the body at the hip joints. In contrast, the muscles of the thigh extend from various points on the pelvis and femur down the entire length of the thigh to the tibia and fibula below the knee, some of them passing over both the hip and knee joints. The muscles that pass over two joints (called *polyarticular* as opposed to *monoarticular* muscles because they cross two joints) perform a special function in large-scale movements, having the ability to act on the knee and hip together in a wide arc of movement such as walking, running,

Fig. 79. The hamstring muscles

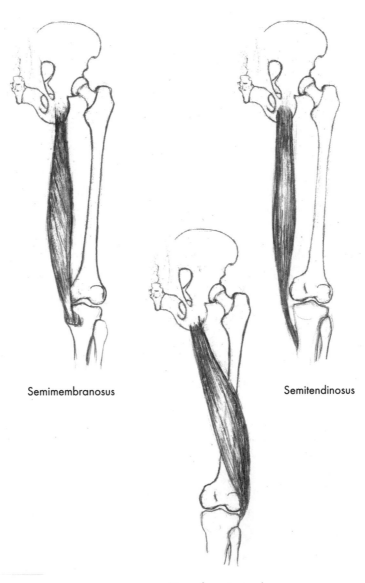

Semimembranosus

Semitendinosus

Biceps femoris muscle

or kicking. Also, because of their length, these muscles are able to contract more than the short muscles of the hip, and so can produce more overall movement than shorter muscles.

Among these long muscles we saw that tensor fasciae latae originates from the anterior iliac crest, or hip bone, and inserts into the iliotibial tract, the band of strong fibrous tissue on the outside of the thigh that inserts into the tibia. Gracilis originates at the pubic bones and travels down the inside of the thigh to the tibia. Sartorius is the long, ribbon-like muscle that originates at the ilium and crosses the leg to the inside of the tibia. Some of the hamstrings are also two-joint muscles which, by acting on the back of the tibia and fibula from their origins on the ischium, also effect large-scale movements of the leg. Finally, rectus femoris, one of the four quadriceps, originates at the ilium and attaches indirectly to the tibia via the knee cap, and so acts on both the hip and knee.

Again, these muscles are responsible for the large, swinging movements of the thigh at the hip and the lower leg at the knee. These movements are coordinated by muscles which act on both joints at once, permitting the powerful, fluid motion of the legs which contrasts with the quicker actions of the shorter, stabilizing muscles.

29. The Knee, Lower Leg, and Ankle

The knee joint is the largest in the body. If you look at the *tibia,* or shin bone, you can see that it is the bone of the lower leg that bears all the weight of the thigh, or femur (Fig. 81). At its top, the tibia is very wide and level, forming almost a platform for the bottom or distal end of the femur to sit upon. The femur at this point is also broad, so that the tibia and femur provide a wide, stable area for the articulations and weight-bearing capacity of the knee joint.

The knee is capable mainly of flexion and extension, and so functions like a hinge joint (Fig. 80). The femoral condyles, or knuckles, at the end of the femur form a kind of rocker shape, fitting into two concave areas on the tibial condyles. There are also cartilaginous discs on the tibia, called *menisci,* which accentuate the depressions that the femoral condyles sit upon. The menisci help to cushion the femur, to distribute weight on the tibia, and to increase the area for absorption of synovial fluid for lubricating the joint. The entire knee joint, including the patella and the condyles of the femur and tibia, is enclosed within a capsular ligament which secretes the synovial fluid.

The *patella* (a Latin word meaning "a small pan") is the final component of the knee. We've seen that the quadriceps muscles converge into a single tendon which inserts into the patella. This tendon is continued below the patella, where it inserts into the tibial tuberosity, so that for all intents and purposes the tendon of the quadriceps muscle passes over the knee and inserts into the tibia. When we flex the leg at the knee, the femur rolls and glides on the tibia, the quadriceps tendon slides in the groove between the femoral condyles like a pulley, and the patella slides in relation to the femur.

The knee, however, is not inherently stable or fitted together like the elbow; ligaments provide its main stability. There are four main ligaments supporting the knee. In the middle of the joint, the *anterior* and *posterior*

Fig. 80. The knee joint

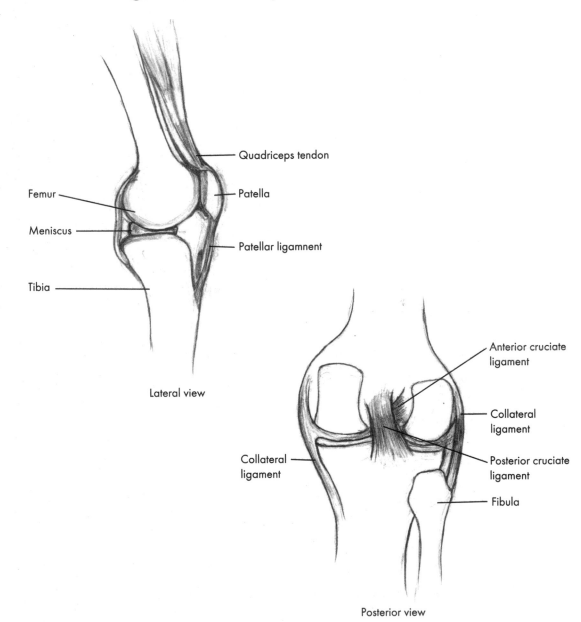

218

Quadriceps tendon

Patella

Femur

Meniscus

Patellar ligamnent

Tibia

Lateral view

Anterior cruciate
ligament

Collateral
ligament

Collateral
ligament

Posterior cruciate
ligament

Fibula

Posterior view

cruciate ligaments "cross," giving these ligaments, which stabilize the knee from front to back, their name. The *collateral ligaments* are on the side and stabilize the joint laterally. When we hyperextend the knee, the ligaments become stretched and the knee "locks." When we flex the knee, most of the ligaments relax and allow for more movement. The knee does rotate slightly during flexion, but for all intents and purposes the knee behaves like a hinge joint that is capable mainly of flexion and extension.

Like the forearm, the lower leg is made up of two bones; these are the *tibia* and *fibula* (Fig. 81). The *tibia,* or shin bone (tibia means "pipe" or "flute"), is quite straight and is the main weight-bearing member of the lower leg. At its upper end, the tibia articulates with the femur to form the knee joint, which as we saw is the largest joint in the body. At its bottom end, the tibia articulates with the *talus* (the ankle bone) to form the hinge joint of the ankle (Fig. 82).

The *fibula* is the second bone of the lower leg. The tibia, which is larger than the fibula, is the main pillar of the lower leg; the fibula runs down the outside of the tibia and articulates with the tibia at its proximal and distal ends to form the *superior* and *inferior tibiofibular joints.* The tibia and fibula are bound together by an *interosseus membrane* just as the ulna and radius are. The fibula is so named because of its resemblance to the Roman fibula, which was a pin used to fasten the toga.

The purpose of the fibula is not as obvious as that of the tibia, but it does in fact have several important functions. In standing, it is the tibia that bears the weight of the femur and distributes this weight to the ankle bones. But the two bones, being bound together by the interosseus membrane, work together to distribute weight and absorb impact from the foot. The meshwork of the interosseus membrane helps the bones of the lower leg to act as a shock absorber, so that although the tibia bears the weight, the two bones operate together to absorb shock, imparting a great deal of flexible strength to the

Fig. 81. Bones of lower leg

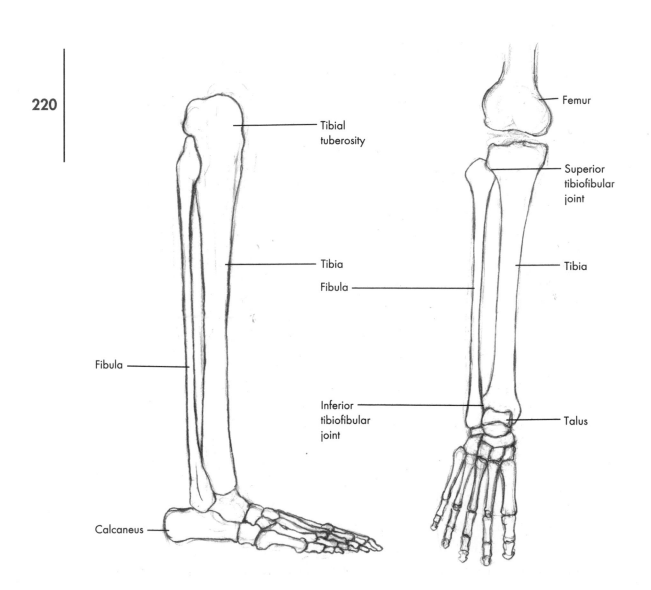

Tibial tuberosity

Tibia

Fibula

Fibula

Calcaneus

Femur

Superior tibiofibular joint

Tibia

Inferior tibiofibular joint

Talus

lower leg, which obviously bears much greater stresses than the forearm. The fibula also provides attachments for muscles. Finally, the two tibiofibular joints, although they do not produce a great deal of movement, and certainly not the supination/pronation seen in the radius and ulna, nevertheless are actively involved in the ankle joint, as we'll see in a moment.

There are a few important landmarks on the lower leg. First of all, the tibia, which is quite large and strong, is a three-sided bone; the shin is the exposed ridge in the front that gets so easily bruised. The bump of bone beneath the knee is the ***tibial tuberosity*** (Fig. 81). When you crawl on your knees, you do not bear weight on the knee cap or knee proper, but on this protrusion at the upper end of the tibia; it corresponds to the olecranon of the elbow, which bears the weight of the upper body when you crawl on elbows and knees. The bump on the outside of the knee is the head of the fibula; you can feel it move when you flex the foot at the ankle.

Let's look now at the ankle joint (Fig. 82). We saw earlier that at its upper end the tibia articulates with the femur to form the knee joint. At its lower end, the tibia sits on the central bone of the ankle, the ***talus,*** which in turn sits on the ***calcaneus,*** the heel bone. So the weight from the femur goes down through the tibia, which sits on the talus and the heel bone; the fibula does not bear any weight at all. Both the tibia and the fibula, however, form the ankle joint. A section of the tibia protrudes down the inside of the talus, called the ***medial malleolus*** (a Latin word meaning "little hammer"). The fibula goes down the outside of the talus (this is called the ***lateral malleolus***). The talus is able to hinge within these bones, allowing dorsiflexion and plantar flexion of the foot—that is, bringing the toes toward the shin, and extending the foot away from the shin. So the bottom end of the tibia sits on top of and articulates with the talus, which is the central ankle bone. And the two malleoli grip the talus on the sides like a pincer, forming a hinge joint. The two malleoli can be felt as the prominent bumps on either side of the ankle, the

Fig. 82. The ankle joint

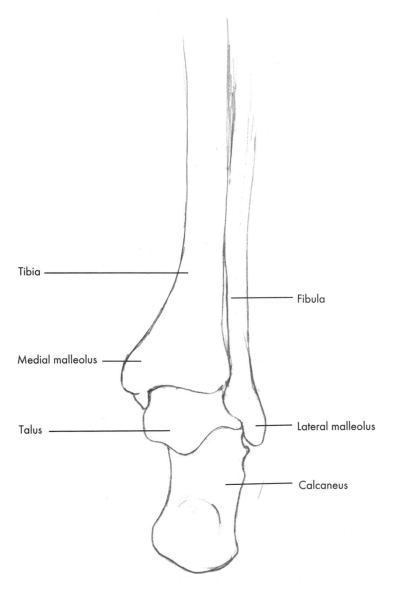

Tibia

Fibula

Medial malleolus

Talus

Lateral malleolus

Calcaneus

Posterior view of ankle

lateral malleolus on the outside (the fibula), and the medial malleolus on the inside (the tibia). If you flex the ankle while holding these two points, you can feel the action of the joint directly in between these "pincers."

Movements of the fibula are also an important component of the ankle joint. We saw earlier that the fibula articulates with the tibia at its two extremities, forming the ***superior tibiofibular*** and ***inferior tibiofibular*** joints (which correspond to the two radio-ulnar joints of the forearm) (Fig. 81). When we flex and extend the ankle, the malleoli must adapt to slight changes in the shape of this joint in order to accommodate movement of the talus within the malleoli. In order for this to happen, these two tibiofibular joints are involved (again, you can feel movement of the fibula at the superior tibiofibular joint if you flex the foot at the ankle). So movement at the joints where the fibula articulates with the tibia is mainly related to movement of the foot at the ankle joint. In this sense, the fibula is like the radius of the forearm: it is important not so much because of its relation to the long bone of the upper part of the limb, but because it is an important component of the ankle joint and the relation of the lower limb to the foot.

There are several ligaments of the ankle worth noting (Fig. 83). On the inside of the ankle, the ***deltoid ligament*** runs from the medial malleolus to the bones of the ankle. On the outside of the ankle, the ***anterior*** and ***posterior talofibular ligaments*** run from the lateral malleolus to the talus, and the ***calcaneofibular ligament*** from the malleolus to the calcaneus. In the common sprain or "turned" ankle, it is the anterior talofibular ligament that typically gets stretched or ruptured.

To complete the bones of the lower leg, there are seven ***tarsal*** or ankle bones ("tarsal" is from the Greek word *tarsus,* which denotes a wickerwork basket used for drying) (Fig. 84). The ***talus*** (Latin for "ankle bone") articulates with the tibia to form the ankle joint. The heel bone, the ***calcaneus*** (from the Latin word *calx,* meaning "the heel") is the largest and strongest of the

Fig. 83. Ligaments of the ankle

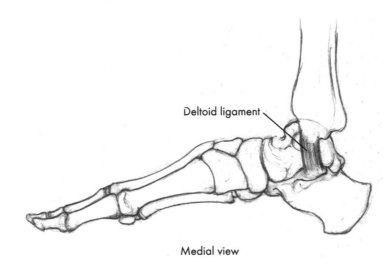

Deltoid ligament

Medial view

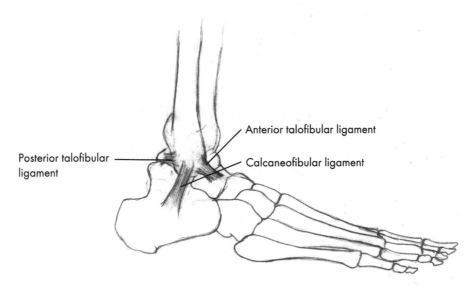

Posterior talofibular ligament

Anterior talofibular ligament

Calcaneofibular ligament

Lateral view

Fig. 84. Bones of foot

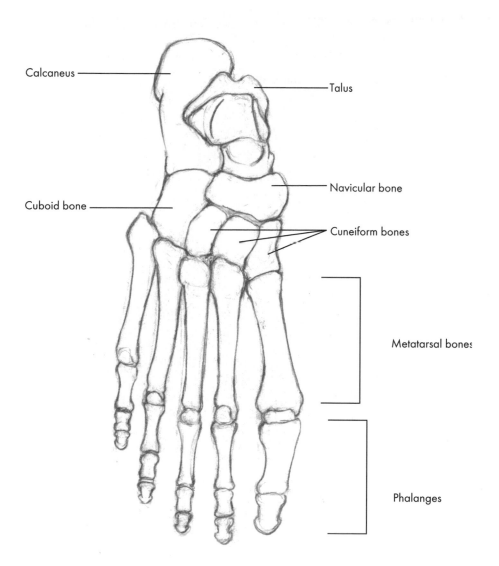

Calcaneus

Talus

Navicular bone

Cuboid bone

Cuneiform bones

Metatarsal bones

Phalanges

tarsal bones. Five more bones make up the tarsals: ***navicular*** (*navicula,* ship), ***cuboid*** (a Greek word meaning "having the shape of a cube"), and three ***cuneiform bones*** (*cuneus,* a wedge + *form*). Then there are five ***metatarsals,*** corresponding to the metacarpals of the hands, and three ***phalanges*** for each little toe and two for the big toe, corresponding to the phalanges of the fingers and thumb. We'll look next at the joints of the foot and the muscles of the lower leg that act on the ankle and the foot.

30. Muscles of the Ankle and Foot

Before discussing the muscles of the lower leg that act on the ankle and foot, let's look at the joints of the foot below the ankle. The ankle joint is the most important, but by no means the only important, joint of the foot, which is actually composed of a number of complex joints. If you could only hinge at the ankle, the foot would not be able to maintain its contact with the ground when confronted with irregularities in terrain or with varying positions of the leg (as when pushing off of one leg). There are several joints in the foot which make it possible to move the foot sideways—in other words, to adjust the foot in relation to the ground. The arches of the foot are also made up of a number of bones (the foot has twenty-six bones in all) which must be able to move in relation to each other in order to give the foot the flexibility it needs to absorb shock.

To talk first about the joints of the foot that make it possible to move the foot sideways: if you move the ankle you'll find that it is possible not simply to hinge at the ankle, but to rotate the ankle freely, as if the ankle is a ball-and-socket joint. As we saw, the talus, which is the keystone of the foot and which forms the ankle joint by articulating with the leg bones, is firmly held within the pincers of the leg bones (the malleoli), and so can only hinge at this point. The other movements of the foot—the sideways motions of inversion and eversion—occur not at the ankle joint, where the talus articulates with the tibia and fibula of the lower leg, but between the talus and the rest of the foot.

There are actually several articulating surfaces which make this possible. First, the two midtarsal bones, the navicular and the cuboid, articulate with the talus and the heel bone, making it possible to move the entire front of the

Fig. 85. Joints of the foot

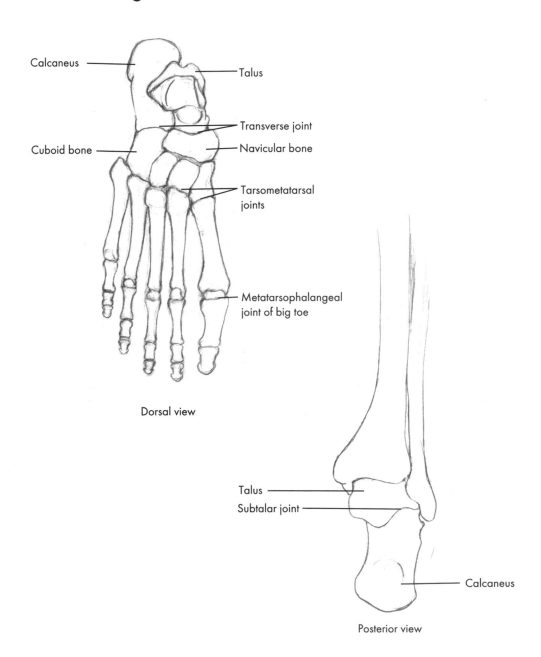

Calcaneus

Talus

Transverse joint

Cuboid bone

Navicular bone

Tarsometatarsal joints

Metatarsophalangeal joint of big toe

Dorsal view

Talus

Subtalar joint

Calcaneus

Posterior view

foot in relation to the talus and heel bones. The navicular bone, which has a depression that fits around the front rounded section of the talus, forms a kind of limited ball-and-socket joint with the talus which makes it possible to rotate the foot around the talus. Also, the cuboid bone is able to move in relation to the heel bone, so that the two midtarsal bones (navicular and cuboid) are able to move in relation to the ankle and heel. This joint, called the ***midtarsal*** or ***transverse*** joint (Fig. 85), is particularly important because most people tend to fix and stiffen the foot with age, when in fact the foot needs to remain mobile and flexible at this joint. If you move the front of the foot back and forth, you can feel how the foot is able to rotate around the talus, and can remain free even when the foot as a whole is bearing weight.

229

There is also a joint where the talus articulates on its underside with calcaneus; this is called the ***talocalcanean,*** or ***subtalar,*** joint (Fig. 85). This joint makes it possible to move the heel bone in relation to the talus bone, and so is part of the articular complex that makes it possible to move the foot in relation to the talus. The important thing to remember about these joints is that they are joints of the foot, not the ankle; as a system, they perform the very crucial function of enabling the foot to be moved with respect to the ground— that is, to turn the foot face in or face out (inversion and eversion). When combined with the hinging of the talus at the ankle, these joints make it possible to orient the foot in every direction, so that the foot seems to operate like a ball-and-socket joint with a great deal of flexibility and mobility.

So the talus has a joint on its top, enabling the foot to hinge in relation to the leg; and it has two joints, in front and below, functioning as one articular complex, which enable the foot to move sideways in relation to the talus and ankle joint. Together these joints—the ankle and the joints of the talus with calcaneus and the midtarsals—give the foot quite a lot of freedom and mobility. In this sense the foot is just like the hand: it has a number of joints— including the tibiofibular joints on the lower leg—which together give it a

Fig. 86. Anterior muscles of the leg

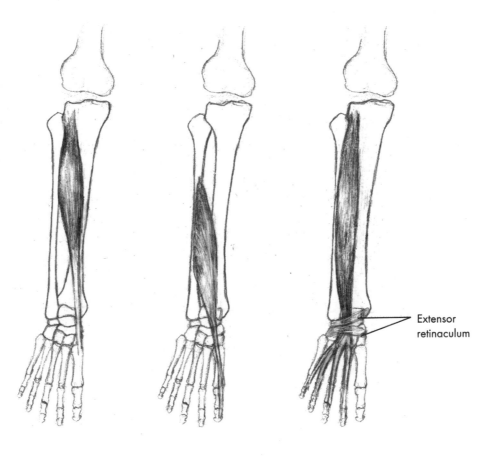

Extensor
retinaculum

Tibialis anterior

Extensor hallucis
longus

Extensor digitorum
longus

great deal of freedom; and it is served by a complex series of muscles in the lower leg, corresponding to the muscles of the forearm, whose tendons pass over the ankle and work the foot and toes.

In addition to the subtalar and transverse joints, there are several other joints in the foot—namely, between the tarsal bones, and between the tarsal bones and the metatarsals (these are called the *tarsometatarsal* joints) (Fig. 85). Since the foot is made up of arches that must adapt to pressure and movement, the various bones of the foot must be able to move slightly in relation to each other. These joints give the foot a great deal of flexibility, so that the arches can act as shock absorbers when walking on uneven terrain and when assuming weight.

Finally, there are the toe joints, which very much resemble those of the hand. As in the hand, there are three in each toe and two in the big toe, and they are all hinge joints except for the ***metatarsophalangeal*** joints, which are capable of adduction and abduction, just like the metacarpophalangeal (knuckle) joints of the fingers. The toes, however, differ from the fingers in several respects: first, the fingers are designed to flex in order to grasp objects and do not extend a great deal, whereas the toes are able to extend significantly in order to accommodate the demands of walking, which requires that the toes bend backward; second, the big toe has lost the ability to oppose the other digits. There are also two small bones, called the ***sesamoid bones,*** located underneath the head of the metatarsal of the big toe, which help to absorb shock (not pictured). The toes are also obviously shorter than the fingers, and not capable of such precise movements. The main joint of the big toe, the first ***metatarsophalangeal joint,*** is particularly crucial for balance and propulsion, because it allows the ball of the foot to be actively pushed off as the leg is advanced. The other joints of the toes are also used in balance, but they are not as crucial as they are in apes, which use them more for prehension.

Fig. 87. Lateral muscles of the leg (peroneal muscles)

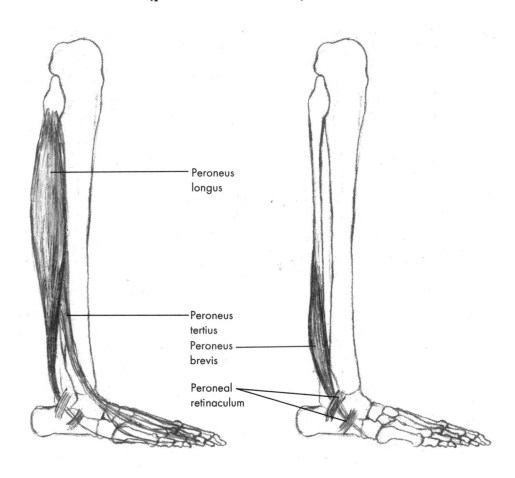

Peroneus
longus

Peroneus
tertius

Peroneus
brevis

Peroneal
retinaculum

Let's now look at the muscles of the lower leg that act on the foot at the ankle. The muscles of this region serve two broad functions. First, they move the foot at the ankle and so help in locomotion. Second, they stabilize the ankle joint when the foot is fixed upon the ground and so are crucial to postural balance and support. As bipeds, we humans are pivoted at the ankle joint, which makes us quite unstable. A four-footed creature is stable both sideways and forward and back; but on two feet, we have no front-to-back base of support. The length of the human foot, which has several points of contact on the ground, along with the foot and ankle muscles, which are highly sensitive to changes in balance, compensate for this front-to-back instability, enabling us to balance on two feet.

As with the hand, a number of muscles act on the foot from the lower leg (the extrinsic muscles of the foot), as well as a number of intrinsic muscles on the foot itself, which we'll look at in the next section. The extrinsic muscles can be divided into three main areas: those on the front of the leg (the extensors); those on the lateral part of the leg, or fibula (the *peroneal* muscles); and those on the posterior leg (the flexors). Like the extrinsic muscles of the hand on the forearm, the muscles coming from these three regions cross the ankle and are held in place by retinacula, or tendinous straps. The tendons from the extensors on the front of the leg pass under the *extensor retinaculum* (which is on the instep and is actually comprised of several bands) and continue on to the dorsum, or top, of the foot (Fig. 86). The *flexor retinaculum* is on the inner, or medial, side of the ankle (not pictured); the tendons from the flexors on the back of the leg cross under this band and under the arch of the foot, and so pass underneath to the sole of the foot. The *peroneal retinaculum* is on the outer side of the foot (Fig. 87); the tendons from the fibular region pass just behind the lateral malleoli, under this retinacular band, to insert into the lateral border of the foot.

Fig. 88. Muscles on the back of the leg

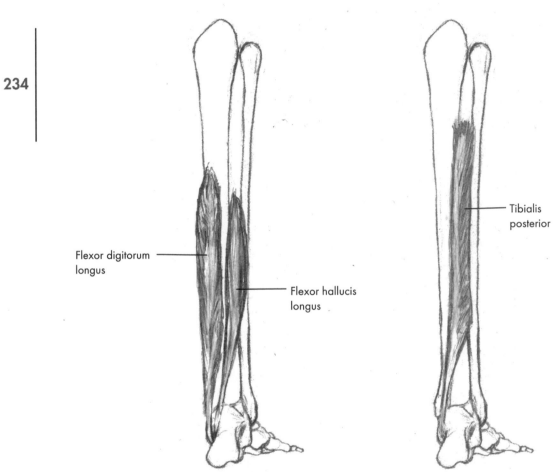

Flexor digitorum
longus

Flexor hallucis
longus

Tibialis
posterior

To look first at the muscles on the front of the leg (Fig. 86), ***tibialis anterior*** arises from the lateral tuberosity and upper shaft of the tibia, as well as the interosseus membrane, and inserts by a long tendon into the cuneiform bone and the first metatarsal. It dorsiflexes and inverts the foot.

Extensor hallucis longus arises from the middle section of the fibula and interosseus membrane and inserts into the last, or distal, phalanx of the big toe. It extends the big toe and flexes the foot.

Extensor digitorum longus, which corresponds to the extensor digitorum of the hand, arises from the shaft of the fibula and the lateral tibial condyle just above the fibula, and from the interosseus membrane. It breaks into four tendons which insert into the tops of the other toes, and dorsiflexes the toes and foot.

There are three peroneal muscles (Fig. 87). ***Peroneus tertius*** is an extension of extensor digitorum longus. It arises from the lower fourth of the fibula and inserts into the fifth metatarsal, or little toe. "Peroneus" simply means "pertaining to the fibula."

There are two lateral muscles of the lower leg, ***peroneus longus*** and ***peroneus brevis.*** Peroneus longus arises from the head and upper two-thirds of the fibula. It inserts by a long tendon, which crosses the sole of the foot obliquely, into the first metatarsal. Peroneus brevis arises from the lower two-thirds of the shaft of the fibula and attaches to the metatarsal of the little toe.

There are five muscles on the posterior region of the leg, or calf, and two behind the knee (Figs. 88 and 89). ***Flexor digitorum longus*** originates from the shaft of the tibia and, splitting into four tendons on the sole of the foot, inserts into the last phalanges of the four smaller toes. This muscle, which helps to form the arch of the foot, flexes the toes and ankle and inverts the foot.

Tibialis posterior, the deepest of the calf muscles, arises from the shaft of both the tibia and fibula and the interosseus membrane and inserts into the navicular and first cuneiform bones. It flexes and inverts the foot; its tendon, along with that of peroneus, forms a sling around both sides of the foot

Fig. 89. Muscles on the back of the leg (cont.)

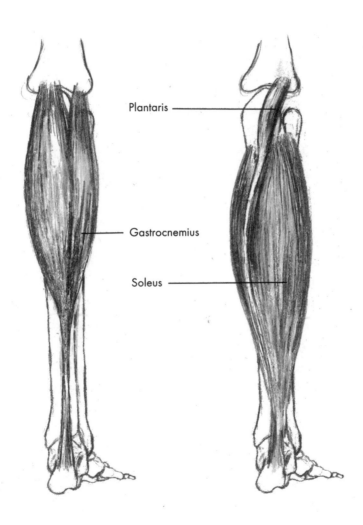

Plantaris

Gastrocnemius

Soleus

that supports the arch of the foot from both sides. Tibialis posterior also serves a crucial postural function. In the same way that the extensors of the knee maintain the extension of the leg at the knee so that we can stand erect, this deep flexor of the leg maintains the perpendicular support of the tibia and fibula above the ankle joint, and in this sense it is a crucial part of the extensor system of muscles that support us against gravity.

Flexor hallucis longus arises from the lower two-thirds of the shaft of the fibula and the interosseus membrane and inserts by a long tendon into the last phalanx of the big toe. ("Hallucis" refers to the big toe, corresponding to "pollicis," which refers to the thumb.) In addition to helping form the arch of the foot, this muscle flexes the toe and ankle and is involved in propelling oneself in walking and running.

The two largest muscles of the calf are *soleus* and *gastrocnemius.* They both insert into the back of the foot via the Achilles tendon, which is the strongest and thickest tendon in the body. This tendon is about six inches long and narrows to the very prominent ridge at the back of the ankle, but widens just below that to insert into the entire lower part of calcaneus. *Soleus,* so named because of its resemblance to the sole-fish, is broad and flat, and lies just underneath gastrocnemius. It arises broadly from the head and upper shaft of the fibula and tibia, converging into the Achilles tendon to insert into calcaneus. It flexes the foot at the ankle, and also acts posturally to maintain the support of the leg in standing. *Gastrocnemius* (*"gastroknemia"* is a Greek word meaning "the belly of the calf," or calf muscle) is the most superficial muscle of the calf and forms most of its bulk. It originates by two heads just above the condyles of the femur and inserts into calcaneus via the Achilles tendon, which it shares with soleus. Its most obvious action is to produce plantar flexion of the foot—that is, to powerfully flex the foot at the ankle. However, in concert with the hamstrings, the gastrocnemius muscle acts as an extensor of the leg at the knee, since it crosses the back of the knee and therefore assists in drawing the knee backward when the leg is flexed at the knee.

Finally, *plantaris* is a small muscle just behind the knee that has a remarkably long tendon running between soleus and gastrocnemius and inserting into the heel. Arising from the shaft of the femur just above the knee, its belly, which is three or four inches long, crosses the back of the knee. Its long tendon obliquely crosses the calf and inserts into calcaneus along with the Achilles tendon. This muscle is an accessory to gastrocnemius.

31. Intrinsic Muscles of the Foot

We looked in the last section at the muscles of the leg that move the foot at the ankle and flex and extend the toes. Let's turn now to the intrinsic muscles of the foot (the muscles on the foot itself), which are quite complex and which assist in the delicate job of balance and propulsion. On the dorsum of the foot there are only two muscles, but on the sole, or plantar surface, of the foot there are well over twenty muscles occurring in four layers—just as complex as those of the hand. ("Plantar" comes from the Latin word meaning "the sole of the foot" and corresponds to the palmar surface of the hand.) Since we are not observing the muscles by performing a dissection in layers, we'll group these muscles according to function just as we did with the intrinsic muscles of the hand.

Extensor digitorum brevis and *extensor hallucis brevis* are the two muscles on the dorsum of the foot (Fig. 90). They originate at the calcaneus and, crossing the foot obliquely, break into four tendons, extensor hallucis brevis inserting into the first phalanx of the big toe, and extensor digitorum brevis inserting into the phalanges of the second, third, and fourth toes. These muscles assist the extrinsic extensors of the toes, extensor hallucis longus and extensor digitorum longus, whose pulls they counteract slightly so that the muscles acting together can evenly extend the toes.

The muscles on the sole of the foot can be divided into four groups: the interossei muscles occupying the spaces in between the metatarsal bones; the muscles that act on the little toe; the muscles that act on the big toe; and the muscles that assist in flexion of the four smaller toes.

The *interossei* muscles of the foot, which correspond to those of the hand, are the deepest layer of muscles on the sole of the foot (Fig. 91). These muscles occupy the spaces between the metatarsal bones, where they originate, and insert into the phalanges of the little toes. There are four *dorsal* and three

Fig. 90. Intrinsic extensors of the foot

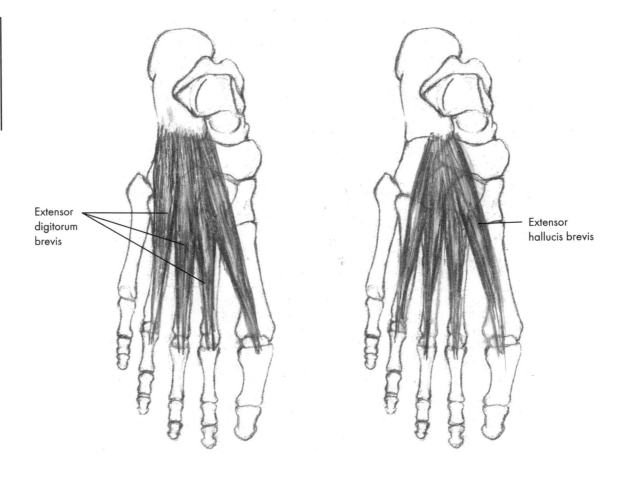

Extensor
digitorum
brevis

Extensor
hallucis brevis

Fig. 91. Interossei muscles

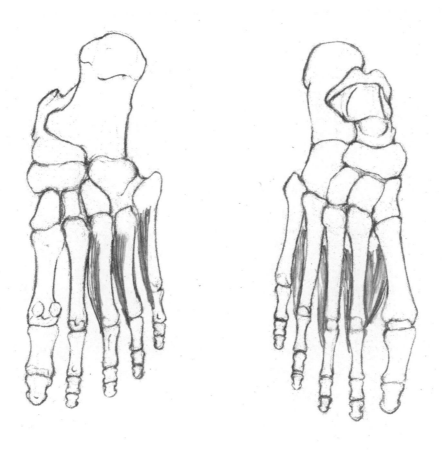

Plantar interossei   Dorsal interossei

Fig. 92. Intrinsic muscles of the little toe

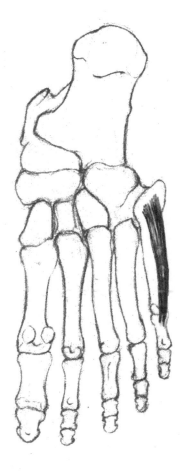

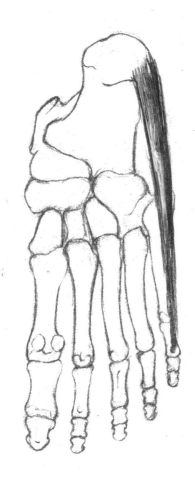

Flexor digiti minimi brevis

Abductor digiti minimi

plantar interossei; the dorsal muscles abduct and also flex the toes, and the palmar interossei adduct the toes.

There are two muscles of the little toe, which lie along the lateral compartment of the foot (Fig. 92). *Abductor digiti minimi,* which abducts the little toe, corresponds to the abductor of the little finger of the hand. It originates at the heel bone (calcaneus) and attaches to the side of the first phalanx of the little toe. *Flexor digiti minimi brevis,* which flexes the little toe, corresponds to the same muscle of the hand. It arises from the metatarsal bone of the little toe and inserts into the first phalanx of the little toe on its outer side.

There are three muscles of the big toe, which lie along the medial compartment of the foot, except for the adductor, which crosses the foot from the lateral side (Fig. 93). *Abductor hallucis* arises from calcaneus and attaches to the side of the first phalanx of the big toe; it abducts and flexes the big toe.

Flexor hallucis brevis arises from the cuboid and cuneiform bones and divides into two sections which insert on either side of the first phalanx of the big toe. This flexor aids in balance.

Adductor hallucis, which adducts the big toe, has two heads, both of which insert into the first phalanx of the big toe. One head arises from the second, third, and fourth metatarsals and crosses the foot obliquely to insert into the toe; the other arises from the ligaments of the third, fourth, and fifth toes and runs across the transverse arch of the foot which, along with tibialis posterior and peroneus longus, it helps to support.

There are several muscles assisting in flexion of the toes (Fig. 94). *Quadratus plantae* (or *flexor accessorius*) originates at the calcaneus and joins the tendon of flexor digitorum longus. As its name suggests, it assists flexor digitorum longus in flexing the toes; it also does so in line with the axis of the toes and so redirects the oblique pull of the long flexor of the toes into a direct backward pull.

The four *lumbrical muscles,* like their counterparts on the hand, arise from the tendon of the extrinsic flexor muscle (flexor digitorum longus) and

Fig. 93. Intrinsic muscles of the big toe

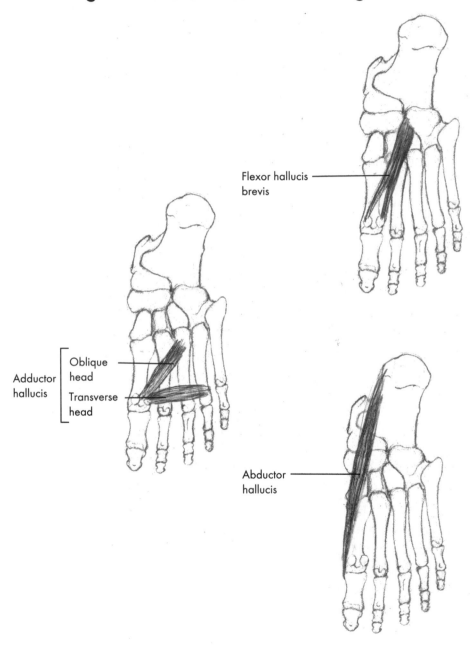

Flexor hallucis brevis

Adductor hallucis

Oblique head

Transverse head

Abductor hallucis

Fig. 94. Intrinsic flexors of the toes

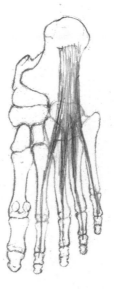

Quadratus plantae

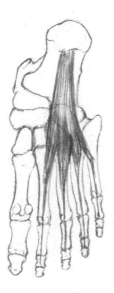

Flexor digitorum
brevis

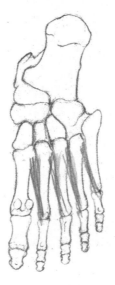

Lumbricales

Fig. 95. Arches of the foot

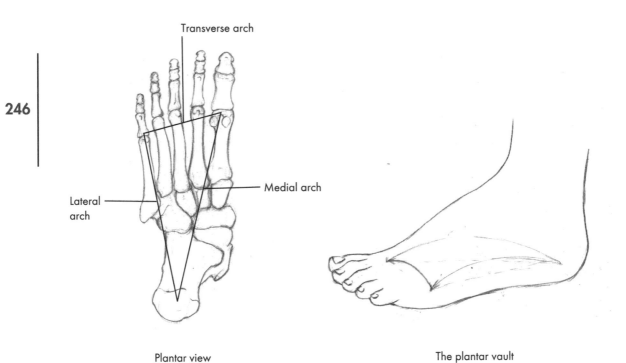

Transverse arch

Medial arch

Lateral
arch

Plantar view

The plantar vault

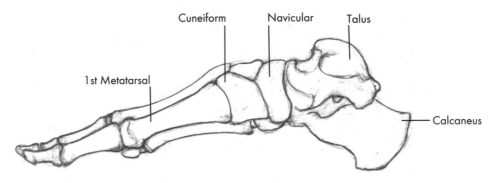

Cuneiform   Navicular   Talus

1st Metatarsal

Calcaneus

The medial arch

insert into the tendons of extensor digitorum longus, which insert into each of the little toes. Along with the interossei, they are flexors of the toes.

Flexor digitorum brevis arises from calcaneus and, splitting into four tendons, inserts into the middle phalange of each of the four little toes. It flexes the middle and first phalanges.

To talk about the arches of the foot, most people think of the foot as having one long arch running between the front of the foot and the heel. However, the foot has not two but three points of support for bearing weight—the ball of the big toe, the ball of the little toe, and the heel—with arches between each of them, forming a three-sided vaulted structure, sometimes called the *plantar vault* (Fig. 95). The main arch is called the *medial arch;* it is easy to feel along the inside of the foot and clearly comes up off the ground. The arch between the ball of the little toe and heel runs down the outside of the foot and is called the *lateral arch* (it isn't so easy to feel because it is filled by muscle). Finally, the arch across the two front points is called the *transverse arch* (also not easy to feel). The arches are easy to understand if you identify the three main weight-bearing areas—the ball of the big toe, the ball of the little toe, and the heel—and draw a triangle between them. The three lines describe the three arches—two longitudinal arches and one transverse arch—which act as an elastic shock absorber and make it possible for the foot to adjust to changing surfaces when walking on uneven ground.

In standing, most of the weight from the tibia, which sits on the talus, is transferred into the heel; but some weight is transferred forward as well. This means that the weight from the lower leg does not come down directly over the heel or directly over the three arches, but toward the back of this long triangle, so that well over half of the body weight is carried on the heel.

The arches of the foot, which are built into its bony structure, are supported by both ligaments and muscles. The medial arch is supported by tibialis posterior, peroneus longus, flexor hallucis longus, flexor digitorum longus, and abductor hallucis. The lateral arch is supported by peroneus longus, peroneus brevis, and abductor digiti minimi, which lie along the outside of the foot. And the transverse arch is supported by peroneus longus, tibialis posterior, and adductor hallucis, which cross the transverse arch. So the arches of the foot are actively supported by muscles and in this sense are part of the postural system, which explains why the condition of the feet and arches is dependent on the overall working of the postural muscles that support the skeleton.

INDEX

A

Abdominal muscles, 125–29, *126, 128*

Abduction, 11

Abductor digiti minimi, 181, *182,* 243

Abductor hallucis, 243, *244*

Abductor pollicis brevis, 179, *180*

Abductor pollicis longus, *176,* 177

Abs. *See* Rectus abdominis muscle

Acetabulum, *188,* 193

Acromioclavicular joint, *144,* 147

Acromion, 143, *144, 145*

Adam's apple, 61

Adduction, 11

Adductor brevis, 207, *208*

Adductor hallucis, 243, *244*

Adductor longus, *208,* 209

Adductor magnus, *208,* 209

Adductor pollicis, 179, *180*

Agonists, 13, 19

Anal canal, *202*

Anatomy

    areas of, 12–13

    comparative, 12

    definition of, 3

    functional, 3–4

    gross, 13

Ankle

    joint, 221, *222,* 223

    ligaments of, 223, *224*

    muscles of, 233, 235, 237

Antagonists, 13, 19

Anterior (direction), 9, *10*

Anterior arch, *78*

Anterior cruciate ligament, 217, *218,* 219

Anterior inferior iliac spine, *190*

Anterior longitudinal ligament, 83, *84*

Anterior sacroiliac ligament, *194,* 195

Anterior superior iliac spine, *190,* 191

Anterior talofibular ligament, 223, *224*

Aponeurotic tissue, 125

Appendicular skeleton, 13

Arches

    foot, *246,* 247–48

    vertebral, 75, *76*

Arm. *See also* Elbow; Hand

    muscles of fore-, 161–66, *162, 164*

    muscles of upper, 155–59, *157, 158*

Arthritis, 16

Articular facets, *116*

Articular surface, 13, *78*

Arytenoid cartilages. *See* Pyramid cartilages

Atlanto-axial joint, *80,* 81

Atlanto-occipital joint, 26, 29, 35, 79, *80*

Atlas, 26, *74, 78,* 79, *80*

Axial skeleton, 13

Axis, *74, 78,* 79, *80,* 81

B

Back muscles

    deep layers, 89–98, *90, 92, 96*

    middle and superficial layers, 99–104, *100, 102, 104*

    summary of, 105–6

Biceps brachii, 153, *157,* 159, 163
Biceps femoris, 213, *214*
Bones, 15. *See also individual bones*
Brachialis, *157,* 159, 165
Brachioradialis, *157,* 159, 165
Breathing. *See* Respiration
Buccinator muscle, *38,* 39, *54,* 57

C
Calcaneofibular ligament, 223, *224*
Calcaneus, *220,* 221, *222,* 223, *225, 228, 246*
Calf. *See* Lower leg
Capitate bone, 167, *168*
Cardiac muscle, 17
Carpal bones, 161, 167, *168*
Carpal tunnel, 166
Cartilage, 15–16
Caudal (direction), 9
Cheek bones, 23
Circumduction, 11
Clavicles, 143, *144, 145*
Coccygeus muscle, 201, *202*
Coccyx (tailbone), 73, *74*
Collar bones. *See* Clavicles
Collateral ligament, *218*
Comparative anatomy, 12
Compressors, 37
Condyle, 14
Constrictors, *32, 54,* 57
Contraction, isotonic vs. isometric, 19
Coracobrachialis, 155, *157*
Coracoid process, *145,* 147
Coronal plane, 7, 29
Corrugator, 37, *38*
Cortical opposition, 12

Costal arch, 113, *114*
Costovertebral joints, 113, *116*
Cranial (direction), 7
Cranium, 23
Crest, 14
Cricoid cartilage, 61, *62, 64*
Cricopharyngeus muscle, *44*
Cricothyroid muscle, *62,* 65
Cuboid bone, *225,* 226, 227, *228,* 229
Cuneiform bones, *225,* 226, *246*

D
Deltoid ligament, 223, *224*
Deltoid muscle, 153, 155, *156*
Dens. *See* Odontoid process
Depressor anguli oris, *38,* 39
Depressor labii inferioris, *38,* 39
Depressors, 37
Diaphragm, *108,* 109, 113, 121, *122,* 123
Diarthroses, 15
Digastric muscle, *30, 33,* 43, *45, 49,* 50
Dilators, 37
Directions, anatomical, 7, 9, *10*
Distal (direction), 9, *10*
Dorsal (direction), 9
Dorsal interossei, *183,* 240, *241,* 243
Dorsiflexion, 12

E
Eating, 57, 59
Elbow, *162,* 163
Embryology, 13
Epicranius muscle, 37, *38*
Epiglottis, *58,* 59
Erector spinae, 95, *96*
Esophagus, *32, 54,* 57, *58*

Ethmoid bone, 23

Eversion, 12

Extension, 11

Extensor carpi radialis brevis, 173, *174, 175*

Extensor carpi radialis longus, 173, *174, 175*

Extensor carpi ulnaris, 173, *174, 175*

Extensor digiti minimi, *176,* 178

Extensor digitorum, *176,* 177

Extensor digitorum brevis, 239, *240*

Extensor digitorum longus, *230,* 235

Extensor hallucis brevis, 239, *240*

Extensor hallucis longus, *230,* 235

Extensor indicis, *176,* 177

Extensor muscles, *28*

Extensor pollicis brevis, *176,* 177

Extensor pollicis longus, *176,* 177

Extensor retinaculum, 165–66, *176, 230,* 233

External abdominal oblique muscle, 125, *126*

External intercostal muscles, *118,* 119

F

Face

    bones of, 23

    muscles of, 37, *38,* 39

Facial expressions, 37, *38,* 39

Fascia, 17

Femur, *192,* 193, 195, *218, 220*

Fibers, 18–19

Fibula, *218,* 219, *220, 222,* 223

Fingers. *See also* Thumb

    bones of, *168,* 169

    extrinsic muscles of, *175, 176,* 177–78

    intrinsic muscles of, 179, 181, *182, 183*

    joints of, 169, 171

Flexion, 9

Flexor accessorius, 243

Flexor carpi radialis, 173, *174, 175*

Flexor carpi ulnaris, 173, *174, 175*

Flexor digiti minimi brevis, 181, *182,* 243

Flexor digitorum brevis, *245,* 247

Flexor digitorum longus, *234,* 235

Flexor digitorum profundus, 173

Flexor digitorum superficialis, 177

Flexor hallucis brevis, 243, *244*

Flexor hallucis longus, *234,* 237

Flexor pollicis brevis, 179, *180*

Flexor pollicis longus, 177

Flexor retinaculum, 166, 233

Foot

    arches of, *246,* 247–48

    bones of, *223, 225,* 226, 227

    extrinsic muscles of, *230, 232,* 233–38, *234, 236*

    intrinsic muscles of, 239, *240–42,* 243, *244–46,* 247–48

    joints of, 227, *228,* 229, 231

Foramen, 14

Foramen magnum, *25,* 26, 27, 29

Forearms, 161–66, *162, 164*

Frontal bone, 23, *24*

Frontalis muscle, 37, *38*

Frontal plane, 7, *8,* 29

Frowning, 37

Functional anatomy, 3–4

Funny bone. *See* Medial epicondyle

G

Gastrocnemius, *236,* 237

Gemellus inferior, *203,* 205

Gemellus superior, *203,* 205

251

Genioglossus muscle, 47, *48*

Geniohyoid muscle, 46, *48, 49,* 50

Glenoid cavity, *145,* 147

Glottis, 65

Gluteus maximus, *204,* 206

Gluteus medius, *204,* 205

Gluteus minimus, *204,* 205

Gracilis, 209, *210,* 215

Greater trochanter, *192,* 195

Gross anatomy, 13

H

Hamate bone, 167, *168*

Hamstrings, 207, 211, 213, *214,* 215

Hand. *See also* Fingers; Thumb

    bones of, 161, 167, *168,* 169

    extrinsic muscles of, 171, 173, *175–76, 177–78*

    intrinsic muscles of, 179–84, *180, 182, 183*

    joints of, 169, 171

Head. *See also* Skull

    balance of, 29, 35, 95

    joints of neck and, *80*

Hip

    joint, 191, *192,* 193, *196,* 197

    muscles of, 201, *203–4,* 205–6

Histology, 13

Horizontal plane, 7

Humerus, *145,* 147, *162*

Hyoglossus muscle, 47, *48*

Hyoid bone, *30,* 31, 33, 47, *48, 62, 80*

Hyperextension, 11

Hypothenar eminence, 179

I

Iliac crest, *188, 190,* 191

Iliacus, 199, *200*

Iliocostalis cervicis, *96,* 97

Iliocostalis lumborum, *96,* 97

Iliocostalis thoracis, *96,* 97

Iliofemoral ligament (Y-ligament), *196,* 197

Iliopsoas muscle, 199, *200*

Iliopubic eminence, 197

Iliotibial tract, 206, 211, 215

Ilium, *188,* 189

Inferior (direction), 9, *10*

Inferior articular process, *76,* 77

Inferior constrictor, *54,* 57

Inferior maxilla, 23

Inferior radio-ulnar joint, 161, *162*

Inferior tibiofibular joint, 219, *220,* 223

Infraspinatus, 153, *154*

Infraspinatus fossa, *145*

Innominate bone, *188,* 189

Insertions, 18

Intercostal muscles, *118,* 119

Internal abdominal oblique muscle, *126,* 127

Internal intercostal muscles, *118,* 119

Interossei muscles, 181, *183,* 239, *241*

Interosseus membrane, 161, 219

Interosseus sacroiliac ligament, 195

Interphalangeal joints, 169, 171, *172*

Interspinalis muscle, 91, *92*

Interspinous ligament, 83, *84*

Intertransverse ligament, 83, *84*

Intertransverse muscle, 91, *92*

Intervertebral discs, *76,* 85

Intervertebral foramina, *76,* 77

Inversion, 12

Involuntary muscle, 17
Ischial tuberosities, 189, *190,* 193
Ischiofemoral ligament, *196,* 197
Ischium, *188,* 189
Isometric contraction, 19
Isotonic contraction, 19

J
Jaw, 35
    bones of, 23
    joint of, *34*
    muscles of, *34, 39, 40,* 41
Joint capsule, 16
Joints, 15–16. *See also individual joints*

K
Kinesiology, 13
Knee, 217, *218,* 219

L
Lachrymal bones, 23
Lamina, 75, *76*
Laryngopharynx, *58*
Larynx, *30, 32,* 33, *58, 62*
    extrinsic (suspensory) muscles of, 43,
        44–45, 46, 61
    intrinsic muscles of, 61–63, *64,* 65–66
Lateral (direction), 9, *10*
Lateral arch, *246,* 247
Lateral cricoarytenoid muscle, 63, *64*
Lateral epicondyle, *162,* 163, 173, *192*
Lateral flexion, 11
Lateral malleolus, 221, *222*
Lateral pterygoid muscle, *40,* 41
Latissimus dorsi, 89, 103, *104,* 135, *146,*
    147
Laughing, 37, 39

Leg. *See* Knee; Lower leg; Thigh
Lesser trochanter, *192,* 195
Levator anguli oris, 37, *38*
Levator ani, 201, *202*
Levator costae muscle, 91, *92,* 135
Levatores costarum muscles, 91, *92,* 135
Levator labii superioris, 37, *38*
Levator labii superioris alaeque nasi, *38*
Levator scapulae, 101, *102, 148,* 149
Levator veli palatini, 51, *52*
Ligaments, 16–17. *See also individual*
    ligaments
Ligamentum flava, 83, *84*
Ligamentum nuchae, 83, 85
Linea alba, 125
Lip, 14
Longissimus capitis, *96,* 97
Longissimus cervicis, *96,* 97
Longissimus thoracis, *96,* 97
Longus capitis, 69, *70,* 107, *108*
Longus colli, 69, *70,* 107, *108*
Lower leg
    bones of, 219, *220,* 221, 223
    landmarks of, 221
    muscles of, *230, 232,* 233–38, *234, 236*
Lumbosacral joint, 195
Lumbrical muscles, 181, *183,* 243, *245*
Lunate bone, 167, *168*

M
Malar, 23
Mandible, 23, *24,* 35, 39
Masseter muscle, *40,* 41
Mastoid process, *24, 25,* 26, 31
Maxilla, 23, *24*
Medial (direction), 9, *10*

Medial arch, *246,* 247

Medial epicondyle, *162,* 163, 171, *192*

Medial malleolus, 221, *222*

Medial pterygoid muscle, *40,* 41

Median plane, 7, *8*

Menisci, 217, *218*

Mentalis muscle, *38,* 39

Metacarpal bones, 161, *168,* 169, *172*

Metacarpophalangeal joint, 171, *172*

Metatarsal bones, *225,* 226, *246*

Metatarsophalangeal joint, *228,* 231

Mid-carpal joint, 169

Middle constrictor, *54,* 57

Midtarsal joint, 229

Monoarticular muscles, 213

Morphology, 13

Mouth, floor of, *49,* 50

Movement, types of, 9, 11–12

Multifidus, 89, *90*

Muscles, 17–19. *See also individual muscles*
    attachments of, 18
    fibers in, 18–19
    forms of, 18–19
    types of, 17

Mylohyoid muscle, 46, *49,* 50

N

Nasal bones, 23

Nasal cavity, *58*

Nasopharynx, *58*

Navicular bone, *225,* 226, 227, *228,* 229, *246*

Neuroanatomy, 13

Nostrils, 37

O

Obliquus capitis inferior, *94,* 95

Obliquus capitis superior, *94,* 95

Obturator externus, *203,* 205

Obturator foramen, *188*

Obturator internus, *203,* 205

Occipital bones, 23, *24, 25, 52*

Occipital condyles, *25,* 26, 29, *78*

Occipital protuberance, *25,* 26, 31

Occiput, 31

Odontoid process (dens), *78,* 81

Olecranon, *162,* 163

Omohyoid muscle, *45,* 46

Ontogeny, 14

Opponens digiti minimi, 181, *182*

Opponens pollicis, 179, *180*

Orbicularis oculi, 37, *38*

Orbicularis oris, *38,* 39, *54,* 57

Origins, 18

Oropharynx, *58*

Os innominatum. *See* Innominate bone

Osteoarthritis, 16

P

Palate, 23, 51–55, *52, 58*

Palatoglossus muscle, 47, 51, *52*

Palatopharyngeus muscle, 51, *52*

Palmar interossei, *183*

Palmaris longus, 173, *174, 175*

Parietal bones, 23, *24*

Patella, *212,* 217, *218*

Patellar ligament, *218*

Pectineus, 207, *208*

Pectoralis major, *150,* 151

Pectoralis minor, 149, *150,* 151

Pedicles, 75, *76*

Pelvic diaphragm, 201, *202*

Pelvis. *See also* Hip
    bones of, *188*
    functions of, 187, 199
    landmarks of, *190,* 191
    ligaments of, *194,* 195, 197
    muscles of, 199, *200,* 201, *202*
    structure of, 187, 191
Peroneal retinaculum, *232,* 233
Peroneus brevis, *232,* 235
Peroneus longus, *232,* 235
Peroneus tertius, *232,* 235
Phalanges, 161, *168,* 169, *225,* 226
Pharyngeal tubercle, *25, 32,* 33, 57
Pharynx. *See* Throat
Phylogeny, 14
Pinched disc, 85, *86*
Piriformis, 201, *203*
Pisiform bone, 167, *168*
Planes, 7, *8*
Plantar flexion, 11
Plantar interossei, *241,* 243
Plantaris, *236,* 238
Plantar vault, *246,* 247
Platysma muscle, *38,* 39
Polyarticular muscles, 213
Posterior (direction), 9, *10*
Posterior cricoarytenoid muscle, 63, *64*
Posterior cruciate ligament, 217, *218,* 219
Posterior inferior iliac spine, *190*
Posterior longitudinal ligament, 83, *84*
Posterior sacroiliac ligament, *194,* 195
Posterior superior iliac spine, *190,* 191
Posterior talofibular ligament, 223, *224*
Posture, 89, 93, 95, 97, 109, 127
Prime movers, 13

Procerus muscle, 37, *38*
Process, 14
Pronation, 12
Pronator quadratus, *164,* 165
Pronator teres, *164,* 165
Proximal (direction), 9, *10*
Psoas major, *108,* 109, 199, *200*
Psoas minor, 109, 199, *200*
Pubic bone, *188,* 189, 193
Pubic symphysis, 189, *190,* 193, *202*
Pubofemoral ligament, *196,* 197
Pyramidalis, 129
Pyramid (arytenoid) cartilages, 61, *62,* 63,
    64

Q
Quadratus femoris, 201, *203*
Quadratus lumborum, *108,* 109, 133, 135
Quadratus plantae, 243, *245*
Quadriceps, 207, 211, *212, 218*

R
Radial deviation, 12
Radiocarpal joint, 169
Radius, 161, *162,* 163
Rectus abdominis muscle, 127, *128,* 133,
    134
Rectus capitis anterior, 69, *70,* 93, 107, *108*
Rectus capitis lateralis, 69, *70,* 93, 107, *108*
Rectus capitis posterior major, *94,* 95
Rectus capitis posterior minor, *94,* 95
Rectus femoris, 211, *212,* 215
Respiration, *117,* 119, 121, *122,* 123
Rheumatoid arthritis, 16
Rhomboid major, 101, *102,* 103, *148,* 149
Rhomboid minor, 101, *102,* 103, *148,* 149

Ribs, 113–19, *114, 117, 118,* 123, 131, 133, 135

Ring cartilage. *See* Cricoid cartilage

Risorius muscle, *38,* 39

ROM, 14

Rotation, 11

Rotator cuff muscles, 153, *154,* 155

Rotatores muscles, *90,* 91

S

Sacroiliac joint, 15, 189, *192*

Sacrospinalis, 95, *96*

Sacrospinous ligament, *194*

Sacrotuberous ligament, *194*

Sacrum, 73, *74, 202*

Sagittal plane, 7, *8*

Salpingopharyngeus muscle, 53

Sartorius, 18, 209, *210,* 215

Scalene muscles, 71, 107, *108,* 123, 131, *132,* 149

Scaphoid bone, 167, *168, 172*

Scapula, 143, *144, 148*

Sciatica, 85, 206

Semimembranosus, 213, *214*

Semispinalis capitis muscle, *90,* 91

Semispinalis cervicis muscle, *90,* 91

Semispinalis thoracis muscle, *90,* 91

Semitendinosus, 213, *214*

Serratus anterior, 149, *150*

Serratus posterior inferior, 99, *100,* 101, 135

Serratus posterior superior, 99, *100,* 101, 135

Sesamoid bones, 231

Shield cartilage. *See* Thyroid cartilage

Shoulder girdle, 143–51, *144–45*

Shoulder muscles, 153–56, *154, 156*

Sit bones, 187, 189, 193

Skeletal muscle, 17, 18

Skeleton
    appendicular, 13
    axial, 13

Skull
    base of, and attachments, 27–35, *28, 32*
    bones of, 23, *24, 25, 26*
    point of balance of, *34*

Slipped disc. *See* Pinched disc

Smooth muscle, 17

Soleus, *236,* 237

Sphenoid bone, 23

Spinal cord, 27, 29, 75

Spinalis cervicis, *96,* 97

Spinalis thoracis, *96,* 97

Spine, 14. *See also* Vertebrae
    anterior muscles of the cervical, 69, *70,* 71
    curves of, 87, 109–10
    movements of, 87
    muscles attaching to the front of, 107–10, *108*
    supporting ligaments of, 83–88, *84*

Spinous process, *76, 77, 84*

Spiral musculature, 137–40, *138*

Splenius capitis, 99, *100,* 101

Splenius cervicis, 99, *100,* 101

Sternoclavicular joint, 143, *144*

Sternocleidomastoid muscle, 26, *28,* 31, 131, 133, *134,* 149

Sternohyoid muscle, *45,* 46

Sternothyroid muscle, 43, *44, 45*

Sternum, 113, *114, 144*

Striated muscle, 17

Styloglossus muscle, 33, 47

Stylohyoid muscle, *30,* 33, 43, *45, 48*

Styloid process, *24, 25, 26,* 31

Stylopharyngeus muscle, *30, 32,* 33, 35, 43, *44, 45, 48, 54*

Subclavius, *150,* 151

Sub-occipital muscles, 69, 93, *94, 95*

Subscapular fossa, *145*

Subscapularis, 153, *154*

Subtalar joint, *228,* 229

Superior (direction), 7, *10*

Superior articular process, *76, 77*

Superior constrictor, *54,* 57

Superior maxilla, 23

Superior radio-ulnar joint, 161, *162*

Superior tibiofibular joint, 219, *220,* 223

Supination, 12

Supinator, *164,* 165

Supraspinatus, 153, *154*

Supraspinatus fossa, *145*

Supraspinous ligament, 83, *84*

Sutures, 23, 26

Swallowing, 57, 59

Synovial fluid, 16

Synovial joints, 15

T

Tailbone. *See* Coccyx

Talocalcanean joint, 229

Talus, 219, *220,* 221, *222,* 223, *225,* 227, *228, 246*

Tarsal bones, 223

Tarsometatarsal joints, *228,* 231

Temporal bones, 23, *24*

Temporalis muscle, *40,* 41

Temporomandibular joint, *34,* 39, 41

Tendons, 17

Tensor fasciae latae, *210,* 211, 215

Tensor veli palatini, *52, 53*

Teres major, *146,* 151

Teres minor, 153, *154*

Terminology

    areas of anatomy, 12–13

    directions and positions, 7, 9, *10*

    learning, 4–5

    planes, 7, *8*

    types of movement, 9, 11–12

Thenar eminence, 179

Thigh muscles, 207–15, *208, 210, 212, 214*

Thorax

    muscles of respiration and, 113–23

    suspensory muscles of, 131–36, *132, 134, 136*

Throat, *32, 54,* 57, *58, 59*

Thumb

    bones of, 167, 169

    extrinsic muscles of, *176,* 177

    intrinsic muscles of, 179, *180,* 181

    joints of, 171, *172*

Thyroarytenoid muscle, *64,* 65

Thyrohyoid muscle, 33, 43, *44, 45*

Thyroid cartilage, 61, *62, 64*

Tibia, 217, *218,* 219, *220, 222*

Tibialis anterior, *230,* 235

Tibialis posterior, *234,* 235, 237

Tibial tuberosity, *220,* 221

TMJ syndrome, 41

    Toes

    bones of, *225,* 226

    joints of, 231

    muscles of, *240–42,* 243, *244–45, 247*

Tongue, 47–50, *48*

Tongue bone. *See* Hyoid bone

Trachea, *30, 32, 54,* 57, *58, 62, 80*

Transversalis. *See* Transversus abdominis muscle

Transverse arch, *246,* 247, 248

Transverse arytenoid muscle, 63, *64*

Transverse joint, *228,* 229

Transverse plane, 7, *8*

Transverse process, *76,* 77, *78, 84*

Transversospinalis muscles, 89, *90*

Transversus abdominis muscle, *126,* 127

Transversus thoracis, *120,* 121

Trapezium bone, 167, *168, 172*

Trapezius, 89, 103, *104, 146,* 147

Trapezoid bone, 167, *168*

Trapezo-metacarpal joint, 171, *172*

Triceps brachii, *158,* 159, 163

Triquetral bone, 167, *168*

Trochlea, 161

Tubercle, 14

Tuberosity, 14

Turbinate, 23

U

Ulna, 161, *162,* 163

Ulnar deviation, 12

Uvula, 51, 53, *58*

Uvular muscle, 51, *52*

V

Vastus intermedius, 211, *212*

Vastus lateralis, 211, *212*

Vastus medialis, 211, *212*

Ventral (direction), 9

Vertebrae, 73–81, *74, 76, 78, 80. See also* Spine

    displacement of, 87–88

    number of, 73

    parts of, 75, 77, 79

Vertebral arch, 75, *76*

Vertebral body, 75, *76*

Vertebral centrum. *See* Vertebral body

Vertebral foramen, 75, *76*

Vocal folds, 61, *62,* 63, *64,* 65

Vocalis muscle, *64,* 65

Vocal mechanism, 33

Voluntary muscle, 17

Vomer, 23

W

Whispering, 63, 65

Windpipe. *See* Trachea

Wrists

    bones of, 161, 167, *168*

    joints of, 169, *170*

    muscles of, 171, 173, *174*

Y

Y-ligament. *See* Iliofemoral ligament

Z

Zygomatic bones, 23, *24*

Zygomaticus major, 37, *38*

Zygomaticus minor, 37, *38*

About the Author

Theodore Dimon, Jr., Ed.D. directs the Dimon School for the Alexander Technique in Cambridge, Massachusetts. He became certified to teach the Alexander Technique at London's Constructive Teaching Centre in 1983, and received both his master's and doctorate degrees in Education from Harvard University. A founding director of the North American Society of Teachers of the Alexander Technique (NASTAT), he has served for the past nine years as President of the Alexander Technique Archives, a non-profit organization that promotes research and scholarship on the Technique. He is the author of *The Undivided Self: Alexander Technique and the Control of Stress*.